An Underwater Guide to Anguilla British West Indies

Dr. Stuart P. Wynne

ISBN: 978-1-5272-4178-7

Cataloguing in Publication Data

Copies can be purchased online at www.stuartwynne.co.uk/anguilla-underwater-guide

Wynne, Dr. Stuart P.

An Underwater Guide to Anguilla British West Indies

250 pages

Foreword

It was not until I was over halfway through this book that the enormity of the task first struck. The diversity of life in the ocean is one thing, but the sheer scale of it is another altogether. I had come up with numerous lists of species to be included in this volume, and no sooner had I finalised it, then the very next day I'd spot a creature that I had not seen here before. The problem is, with such extensive potential habitat, uncommon species may go unnoticed for years until one day they are stumbled upon by a lucky explorer, or more often, they themselves stumble upon a fish pot and are inadvertently hauled to the surface a few days later. Not only this, but so many of the creatures that live in the ocean release their hatchling larvae into the open-water where they can drift for hundreds of miles on prevailing currents until they eventually settle, usually by chance, on a suitable habitat. This means, that a particular species can potentially be absent from an area of reef one year, but due to favourable circumstances suddenly appear again the next. This method of larval dispersion can be highlighted by the invasion of the Indo-Pacific Lionfish (*Pterois volitans*) that has been spreading eastwards across the Caribbean on ocean currents since its accidental release in Florida early during the 1990's (see page 178-179). What all this means is that a book such as this will never be able to claim it contains all species that it is possible to find in a given area. The diversity is that great. For example, only today, while contemplating writing this foreword, a fisherman came into my office with a photograph of a creature he had just found in one of his fish traps. In all his years of fishing neither he nor his friends had ever seen this creature before. It turned out to be a Red Banded Lobster (*Justitia longimanus*), a species that no one I have spoken to this afternoon has ever seen here before either. It seems the gods are laughing at me, as moments earlier, I had just decided that my species list was complete (apart from the myriad of small cryptic invertebrates that live out there) because no new species reports had been made to me over recent months. I know now that this will probably never be achieved, so to pay my respects to the majesty of nature I have decided not to add an entry for *J. longimanus* leaving it, instead, as the inspiration behind this foreword. All I hope is that this work will be a valuable resource to both the people of Anguilla and those who visit it, providing them a small insight into the wonderfully fascinating world that exists below the waves.

Dr. Stuart P. Wynne

(written on the 23rd February 2010 after completeing the first draft of book text)

Acknowledgements

Countless texts and internet resources have been used while researching this book, and as such it would be almost impossible to list them all. By far the most important work was that conducted by Paul Humann and Ned Deloach – their publications, known collectively as The Reef Set, have been invaluable. Without these texts this current volume would not have been possible, or at best would have been lacking many of the details that it currently contains. Other reference materials used have been listed towards the end of Chapter 6. Every effort has been made to ensure this volume is up to date and consistent with accepted science, and identifications are as accurate as possible.

Not only have many works of reference been vitally important when producing this book, but so has the help of many individuals either via email or through personal contact. Special thanks should be given to the previous Director of Fisheries in Anguilla, Mr James Gumbs, who supported this project from the onset. Without his support this work would not have been possible. Secondly thanks must go to those, also including James, who proof read various stages of the text in an attempt to remove as many typographical errors as possible: Professor Robin South, Sarah Ramsden, Jane Emery, Farah Mukhida, Steve Donahue & Delene Guenther all dedicated huge amounts of time wading through reams of text, and their constructive comments were greatly valued. Special appreciation also goes to my wife, Marnie Wynne, who not only proof read this book, but was also there from beginning to end, listening to me during the highs and lows of this enormous task, and was supportive and encouraging to the end. Finally a huge thank you goes to Posa Skelton who donated endless hours to the early layout stages of this work and is responsible for such a beautiful final product.

All photos have been credited in the text, except those used as background inserts or other design elements. All of these images were taken by the author except that on page 29, which was taken by Ellen Muller, and page 106, which was taken by Florent Charpin. All images have been used with permission from the owners, and should not be reproduced without their permission. Special thanks goes to Paul Humann, Florent Charpin, Lippold Hakken and Ellen Muller for kindly giving permission for their images to be used despite never having met me, and to two keen divers with a passion for photography who were living in Anguilla at the time, Gert-Jan Vandenbos and Steve Donahue.

List of photographic contributors:

Table of Contents

Chapter 5 - Artifacts   209

Chapter 6 - Supplemental Information   220

Index of Common Names   232

"The sea, once it casts its spell, holds you in its net of wonder forever"

- Jacques Cousteau -

Chapter 1 - Introduction

1.1 Anguilla

Anguilla, located in the Eastern Caribbean, is the most northerly of the Leeward Islands within the Lesser Antilles (figure 1.1). It is a small, low-lying, primarily coralline archipelago consisting of mainland Anguilla and a collection of small offshore cays. The capital, known as The Valley, is positioned centrally on the mainland, being located at approximately 18°12.80 N and 63°03.00 W. The main industries and biggest employers on the island are tourism and offshore financial services, with fishing a relatively small contributor, estimated to provide less than 2% of the GDP. Even at this low level however, because demand is often driven by the expanding tourism market and local population, fishery resources are thought to be under threat, an issue that will be discussed in greater detail in Chapter 6 (page 220-230). The local population was established by census in 2001 as a little less than 12,000, although the actual population is somewhat larger than this due to a growing number of expatriates.

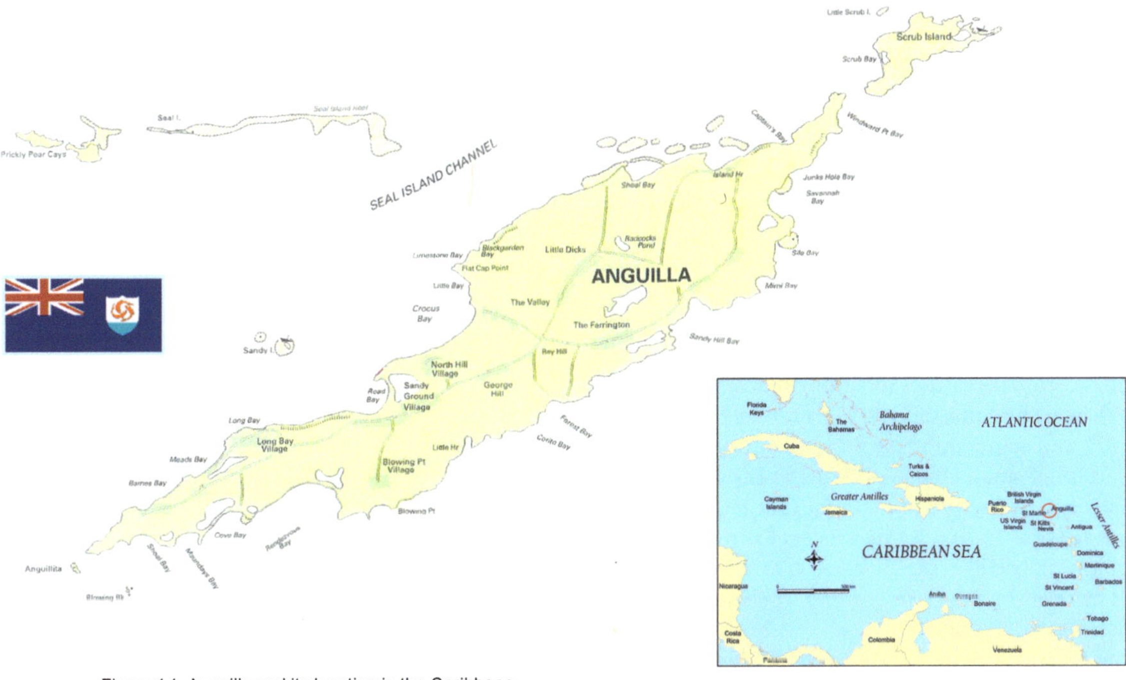

Figure 1.1. Anguilla and its location in the Caribbean

Anguilla is surrounded by a mixture of patch, barrier and fringing reefs interspersed with seagrass beds, sand channels and low relief pavement reef. Aside from many small emergent rocks there are a number of mid-sized cays that support a varying amount of terrestrial life: from the relatively large Dog and Scrub Islands; to the mid-sized Prickly Pear Cays; the small and almost barren Anguillita and Blowing Rock; and the low-lying Seal Island. The majority of these are surrounded with a rich and diverse ocean community, and form the basis for six of the seven Marine Parks that are currently in existence. Five of these Marine Parks primarily consist of coral reef or hard bottom communities: Shoal Bay – Island Harbour, Dog Island, Prickly Pear & Seal Island Reef, Sandy Island, and the newly legislated Sombrero Island. The other two are Little Bay which is an important seagrass area and Junks Hole which is of historical relevance but otherwise of little interest. Rendezvous Bay, although not a Marine Park, is a specially managed seagrass area where anchoring is only permitted by special arrangement. Aside from Junks Hole the Marine Parks are highly important areas for visitors to the island and locals alike, providing important recreational possibilities and being a reliable source of food. Currently the only user restriction that exists in the parks relates to anchoring, with fishing still permitted throughout by residents. This means these areas are under increasing pressure as the tourism industry and local population continue to expand. In actual fact, in a curious twist of fate, until sufficient protective management is introduced, these 'protected' areas are actually under more pressure than non-Marine Park regions, as they attract more visitors and so human-related stresses are higher here than elsewhere.

It is hoped that this book will not only inspire those interested in the undersea world around this beautiful island, but also encourage them to treat it with the necessary respect. It aims to highlight the important work being conducted by various people and organisations on Anguilla to conserve and protect its natural environment and further inspire readers to support them and help preserve the marine world for future generations to come.

1.2 Marine Habitats of Anguilla

Of the many different marine habitats around Anguilla only three will be regularly visited by divers and snorkellers. These are sand flats, seagrass beds and coral reefs. Other marine habitats that may be seen on occasion include mangrove forests, whose roots are frequently submerged, and algal flats. Mangroves have a limited distribution in Anguilla, but provide important nurseries for juvenile fish that later migrate to reef areas. Algal flats are more common but usually occur in deeper regions and this, combined with their relatively low biodiversity, means they are rarely visited recreationally .

Sand Flats

Despite the barren appearance of sand flats, they do in fact support a great variety of life. The majority of creatures that live here are found under the surface of the sand (infaunal), and consist of a myriad of burrowing and filter feeding creatures including many bivalves (page 92-93); heart urchins, sand dollars and sea biscuits (page 101-102); and an uncountable number of worms, crustaceans, and other small creatures. Most of these animals live off the organic portion of material that has drifted down to the sea bed. This recycling process forms an essential part of the nutrient cycle as these creatures are themselves consumed by larger organisms, including a great many fish and ray species.

Other creatures live an epifaunal (on or above the surface) existence, normally feeding on the infaunal community. To camouflage themselves from potential predators, Stingrays (page 202) often lay half-buried in sand, but when active they swim over the sea bed browsing for food buried beneath them. These rays are often accompanied by other opportunistic fish who feed alongside them, for example goatfish (page 194-195) and Bar Jacks (page 110). Other species also cruise the sand flats, such as the Flying Gurnard (page 180), although they are more of a rarity.

Offering an oasis from the seemingly endless sea of sand, small rocks are sometimes present that a limited yet diverse community lives upon. This community ranges from sessile permanent inhabitants like sponges (page 30-38) to transient fish species that move between locations while they continue their endless search for food.

View of Scilly Cay in Shoal Bay - Island Harbour Marine Park

Anguilla's marine life provides tourism opportunities

Smooth Flower Coral (page 63), common in Anguillian waters

Seagrass Beds

Seagrass beds are similar in many ways to sand flats in that a great majority of the creatures that live there are infaunal, although the general appearance is more diverse due to the plants and algae growing on the surface. Seagrasses are the only flowering plants that live underwater in Anguilla. The most common species here is Turtle Grass, although it is often interspersed with Manatee Grass (page 15), and a variety of green algae, for example: calcareous algae, bristle brush algae, and paddle algae - all of which can be found in section 2.3 (page 19-24).

Seagrass beds provide an important habitat for juvenile fish that later migrate to coral reef areas or into deeper waters. At this early stage in their lifecycle they are often overlooked because of their small size and cryptic nature. When adults, some of these juveniles grow up to become important fishery resources, for example the Yellowtail Snapper (page 132). Seagrass also provides a food source for Green Turtles (page 207), and in Florida even manages to sustain populations of Manatees (also known as Dugongs or Sea Cows), a large marine mammal that feeds exclusively on these plants.

Seagrass beds also help to stabilise sandy bottoms through their extensive root systems. These roots hold material in place that might otherwise move during a storm or other rough conditions. Areas dominated by seagrass are common in Anguilla, and include Little Bay, Road Bay, Rendezvous Bay, Forest Bay, Shoal Bay East, Savannah Bay and Little Harbour.

Coral Reefs

Coral Reefs are only rivalled by rain forests in terms of diversity and as such are the richest habitat in Anguillian waters, with a huge number of organisms from many different taxonomic groups present. The majority of species described in this book are associated with coral reefs, some even being responsible for its creation - the so-called 'hard coral' polyps (page 51-63) live in colonies and build a solid calcium carbonate skeleton that ultimately forms the main reef structure. This diversity of life combined with the beauty of reef topography is the main reason divers and snorkellers favour this particular habitat over others. Ship wrecks offer similar opportunities, but are usually too deep to snorkel.

Lush seagrass beds in Island Harbour

Large colony of Finger Coral (page 52) at Sandy Island

Mooring buoys help protect coral reefs

As well as providing a home for an array of species (and thus being an important food source), coral reefs also protect the coastal zone from storm damage by dissipating wave energy before it reaches the shore. If this did not happen a great many of Anguilla's beaches would potentially erode away, which would be an economic disaster because of their importance to the booming tourist industry.

The main reef area in Anguilla stretches from Prickly Pear Cays to Island Harbour, and back along the coast from Island Harbour to Shoal Bay East. This reef is a mixture of barrier and patch reef interspersed with sandy areas. Between Shoal Bay East and Limestone Bay there is a long fringing reef that houses a number of dive sites (see figure 1.2). On the southern side of Anguilla fringing reefs dominate the coastline although much of these areas are in poor condition after failing to recover from storm damage over the past few decades.

Mooring buoys in Little Bay Marine Park

1.3 Diving and Snorkelling in Anguilla

Diving and snorkelling are popular pastimes in Anguilla. Currently there are over 20 active dive sites and a number of popular snorkelling spots (see figure 1.2), although because dive sites are often situated in deeper locations these two activities usually take place in different areas. Visitors to the island are by law only allowed to SCUBA dive in Anguilla if they use a licensed Anguillian dive operator. This is because these dive operators pay a fee to the Department of Fisheries and Marine Resources who coordinate and supply the equipment that is used to install and maintain the dive moorings. These moorings have been put in place primarily to protect the reef from dive boats anchoring and damaging it, and also to facilitate the diving process which is easier if the operator's boat can be moored. At sites that are used for drift dives, or at those locations with a fairly shallow sand patch that can be used for anchoring, no moorings are provided. A brief description of each dive site or snorkelling location is given on pages 9-12.

Snorkelling is a far more accessible sport than diving as all one needs is a mask, snorkel, pair of fins and the sea. Even though one can snorkel practically anywhere there is water, to get the best out of the experience it is worth asking local advice before proceeding, as some areas are far more interesting and easier to reach than others. Although snorkelling is a safe pastime, it is always wise to go with at least one other person, and to let somebody

Making a stand - a young mangrove anchors itself close to the shore

Coral reef community near Black Garden Bay

Coral reef community in Shoal Bay East Marine Park

Dense sea plume forest near Limestone Bay

else know exactly where one is exploring in case any difficulties occur. It is also very important to always know ones own limits, as exceeding them is probably the only real source of danger that this activity poses.

Although some people consider diving a bit more daunting than snorkelling, if properly trained it is also a safe pastime. Again, apart from never diving alone and always making sure somebody else knows exactly where you are going, the best advice is to know your limits and never exceed them. Both diving and snorkelling offer amazing rewards in terms of witnessing visually stunning locations that are otherwise inaccessible to humans. It is hoped that this book will act as a key to unlock the secret world that exists just under the surface by helping explorers identify the inhabitants of this fascinating realm - whether plant, animal, or in some cases, mineral.

1.4 How to use this Book

This book has been divided into three main sections with a couple of supplementary chapters on closely related topics. There is a glossary and an index at the back of this book that are useful to define terminology and allow quick and easy referencing. The table of contents before Chapter 1 is useful for this purpose also.

In order to identify an organism seen while snorkelling or diving the first question to ask is 'Does it look like a plant or an animal?'. If you answered 'plant' to this question then refer to Chapter 2 (page 13-28) as all common plants and algae are detailed there. Be careful not to mistake certain corals for plants (or rocks for that matter!) as they can look quite similar. Also, some corals can look like other invertebrates, for example White Encrusting Zoanthids (page 50). If you answered 'animal' to this question then it is time to ask yourself a second question: *'Does it look like a vertebrate (fish, mammal, reptile) or an invertebrate (worm, mollusc, crustacean, sponge etc)?'*. In most cases the answer to this question should be fairly obvious, although at times it might require the reader to browse the first few paragraphs of Chapter 3 (invertebrates - page 29-105) and Chapter 4 (vertebrates - page 106-208) to be sure of positive identification.

Once the chapter that contains your mystery organism has been identified the real work begins. Each chapter has been organised slightly differently to suit the various groups. For example, the vertebrates that make up the largest chapter in this book have been split into three subgroups: the fish, the mammals, and the reptiles. In the ocean the fish are the largest group so these have been dealt with first. This is as far as assumptive identification can go really, so now you have to put in a little bit of work: a few subjective decisions are needed to identify your creature, which in this example you have decided was a fish. Firstly, think back to its body shape and decide how much it resembled what might be considered a 'classic' fish. The groups within this chapter are divided up based primarily on body type, although later it becomes more to do with basic behaviour as body plans become somewhat unfamiliar and at times downright bizarre! In short, Chapter 4 starts with silvery, simple-looking fish, and generally moves onto the more elaborate designs as it progresses. Once the group has been decided upon (for example 'silver swimmers') it is now a matter of following the steps as described in the 'Notes on identification' section below. Similar notes accompany each chapter, and once you have gone through the identification process a few times it will become less troublesome to find your quarry. As a final pointer remember that nature is wonderfully diverse and so it is very difficult to categorise precisely without 'grey areas'. For this reason always check alternative chapters before deciding the organism you sighted isn't in the book!

Other chapters are structured slightly differently depending on what arrangement best suits the group as a whole. For example Chapter 2 is only divided into quite general large groups, based mainly on pigmentation. Chapter 3 on the other hand is more variable, making it difficult to follow either of these layouts. Instead a mixture is used, sometimes adopting large categorisations to split the groups (as in Chapter 2), but at other times breaking things down into smaller groups (as in Chapter 4).

Although every effort has been made to include all the organisms likely to be seen in the waters around Anguilla, it is not possible to produce a definitive list because of the sheer volume of rare or highly cryptic species. Furthermore some of these organisms are hard to identify, or grow in such a variety of physical forms that it is not practical to include everything. Having said this, if readers can confirm the presence of a creature that is not featured in this book with a good quality photograph, please contact the author so that any subsequent editions of this book may be kept up to date. Full acknowledgement will be given to all who contribute.

1.5 Notes on Identification

There are a number of key features to look out for when identifying a creature, but there are also a number of other aspects that although not key, will be of great help when using this book. For example, when identifying a certain

The Department of Fisheries monitors the health of the marine ecosystem.

Countless creatures adapt to their environment such as this file fish blending in with the segments of a *Halimeda* alga.

Fisheries officers installing mooring buoy anchors

species of fish, although there may be many features that could be described, the aspects that make it unique are generally those described first, together with overall body colour. These features are usually enough when combined with a photograph to make correct identification. However, some species can look very similar to another completely different species (for example, the Fairy Basslet – page 153 and the juvenile Spanish Hogfish - page 161). Here a key identification feature is not body colour, and as they are both small, body features are hard to distinguish. The key here is that their behaviour is very different, with the Fairy Basslet hanging upside-down and almost motionless under ledges, whereas the juvenile Spanish Hogfish swims around actively and acts as a cleaner fish, pecking parasites off the skin of other larger fish. Therefore these behaviours are often mentioned to aid the identification process. Other examples where this can help are: Yellowtail Snapper (page 132) and Yellow Goatfish (page 195); Blue Chromis (page 143) and juvenile Creole Wrasse (page 162). It is worth taking a moment here to refer to these sections and practice your differentiation skills. Other than behaviour, the habitat where a species is found can also be a clue to identification and so that is usually mentioned too.

Other keys to identification are the abundance ratings and maximum sizes that have been assigned to most of the featured organisms. These abundance ratings have been established through ten years of research around Anguilla by the Department of Fisheries and Marine Resources. The rating however is dependent on suitable available habitat and so can vary from site to site. These categories have been divided as described below:

Abundant - Will be seen in large numbers on almost every dive

Common - Will probably be seen more than once during a 30 minute dive

Occasional - Will probably be seen once during a 60 minute dive

Uncommon - Unlikely to be seen, but is sighted every now and again

Rare - Only seen or reported once or twice during the ten year study period

So, this category can be used to narrow down a species that is classed as abundant or common, and rule out those that have been listed as uncommon or rare. Classifications such as this are not without their problems however, as a species

Prickly Pear Marine Park

Little Bay Marine Park

An emergent reef

common in one particular area might be rare in others. In these cases descriptions are given such as 'locally abundant', although it is difficult to be all inclusive in this way as abundances may be seasonal or naturally vary from year to year.

The maximum size has been included as it too can be a useful identification feature at times. It must be appreciated however that a young individual will be much smaller than an older one, and that a maximum size is the size the species would attain if it died of old age. As fishers or other predators usually catch individuals well before they reach their maximum size, a good general rule is that a species will rarely be sighted much larger than half the size stated.

1.6 Notes on Names

The common names that are used in this book are those set down by Humann and Deloach in their Reef Set books. These names are those generally accepted as common names throughout Florida, the Bahamas and wider Caribbean. However, local names exist for many species (especially those caught by fishers), and these have been included where applicable. Local names vary from island to island, hence the need for an accepted common name, although the local names are important to learn, both to preserve local tradition and to enable conversations with fishers that avoid unnecessary confusion! The scientific names assigned to each species are global standards set by the scientific community. They are particularly useful to avoid misunderstandings when a species has a range that covers many regions and as such may have many different common names. These names also help to describe the species' taxonomy but they can change from time to time when taxonomists decide a species has been misclassified. In terms of the capitalisation of names, the standard used within this book is to capitalise common names when they refer to a specific species, such as the 'Red Eye Sponge Crab', but not where the common name refers to a group, in this case the 'sponge crabs'. When using scientific names, the scientific standard has been followed as much as possible, so 'Cetacean' would be capitalised, but its common (group) name 'marine mammal' would not. In the index of common names (page 231) both species and groups have been included for ease of reference. Pages cited in the index are for the main entry within the text where the species or group features. These indexed entries in the main text have been given a semi-bold font.

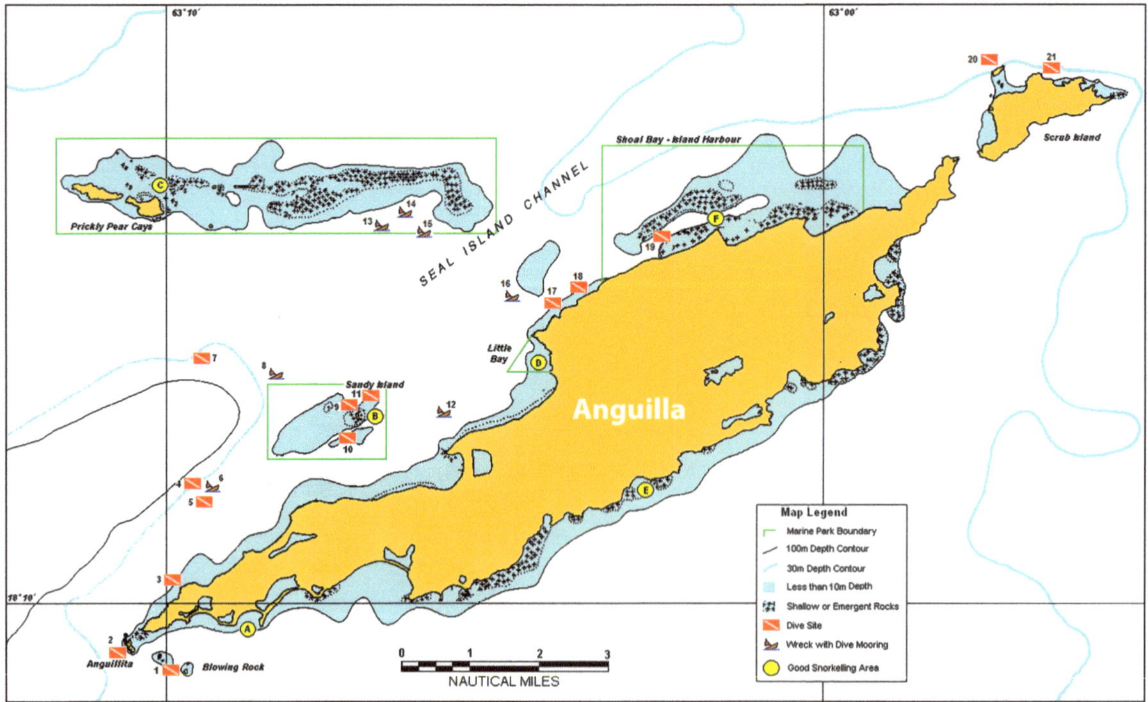

Figure 1.2 Map of Anguilla showing Marine Parks, active dive sites, and suggested snorkelling areas

1.7 Marine Sites of Interest

Figure 1.2 illustrates a number of important marine features that relate to this book. The locations for four of the seven Marine Parks are of main interest (Dog Island & Sombrero Island have been omitted due to their distance from the mainland, and Junks Hole is not shown as diving is prohibited to protect the site from potential looters). Apart from specially designated sandy areas the Marine Parks are "No Anchoring Zones", with mooring buoys installed for charter boat convenience. Further information on the terms and conditions of buoy usage can be found in the Anguilla Marine Park System brochure available at the Department of Fisheries and Marine Resources office, or the Marine Base at Sandy Ground. These brochures are also distributed around the island by the Anguilla National Trust, the Anguilla Tourist Association, local dive operators and a number of other outlets.

Also detailed in figure 1.2 are the main active dive sites in Anguillian waters (not including those at Dog Island). Many of these dive sites have permanent red dive mooring buoys which are for use by licenced Anguillian dive operators only. These moorings are maintained by the Department of Fisheries and Marine Resources, with collaboration from local dive operators, although adverse sea conditions and/or collisions with larger charter boats mean they periodically go missing.

Descriptions of all sites can be found below. Sombrero Island also offers nice diving, but no official site currently exists there. Historically there are a number of other dive sites that were once popular but are now considered disused. These sites are sometimes mentioned in outdated tourist publications and so have been given a brief mention towards the end of the section. Finally, suggested snorkelling areas have been described, the locations of which have also been illustrated in Figure 1.2.

1.8 Dive Sites of Anguilla

These descriptions are based on observations, old documents and personal communications with the various dive operators on Anguilla. Maximum depths and experience levels are based on normal dive conditions, but may vary so use this only as a guide. Adverse weather and strong currents increase difficulty level. Always dive with a licensed Anguillian dive operator.

Bow view of the M.V. Oosterdiep (No.6 Overleaf)

On Map:

1. **Blowing Rock:** Maximum depth 10 m. Experience level – Beginner. Generally shallow with some impressive coral formations in places. This site does not have a mooring, as it is normally used as a drift dive which partially encircles Blowing Rock through impressive underwater rock formations and shoals of glittering fish.

2. **Anguillita:** Maximum depth 18 m. Experience level – Beginner. Starting off in the shallows this dive takes you along a wall with impressive overhangs and many large species of shoaling fish. As the wall comes to an end the dive swings over a reef and back onto a second wall that returns to the mooring. There are a number of small caves to look out for on your way as well as tarpon, stingrays and sharks

3. **Blollyham Bay:** Maximum depth 16 m. Experience level – Beginner. The dive begins offshore at the eastern end of the bay where a gentle current carries you to the west. There are some interesting coral and wall formations present as well as numerous stingrays and occasional rare fish sightings. This site was also previously known as Frenchmans West.

4. **Paintcan Reef:** Maximum depth 20 m. Experience level – Beginner. Although still classed as an active dive site, Diep South Reef (next), a similar and close by site, is usually visited instead because it is in slightly better condition.

5. **Diep South:** Maximum depth 22 m. Experience level – Intermediate. A moderately deep reef with medium relief located south of the M.V. Oosterdiep. This dive takes you on a classic tour of Anguilla's underwater life with some nice coral formations and the chance to see many fish species and an array of invertebrates.

6. **M.V. Oosterdiep** (Wreck; Good Condition): Maximum depth 22 m. Experience level – Intermediate. Home to a number of very friendly turtles, this wreck is over 50 metres long, upright and almost intact although recent storms have started to take their toll. There are, however, still plenty of swim-throughs and other places to explore. Good for large pelagic fish species.

7. **Authors Deep**: Maximum depth 30 m. Experience level – Advanced. Popular for its depth and great visibility. This is the perfect dive site to observe deeper reef species that are not usually found on the shallower sites closer to shore. It is one of the few locations where black coral has been seen.

8. **M.V. Ida Maria** (Wreck; Poor Condition): Maximum depth 18 m. Experience level – Intermediate. Once an impressive wreck 25 metres long, the stern is still upright but the bow is twisted and lies on its side. This damage was caused by recent storms. Having said this, it is still worth a visit as it can attract schools of Barracuda and other large deeper water pelagic species.

9. **No Name Reef**: Maximum depth 18 m. Experience level – Beginner. Described by some as the best of the Sandy Island dive sites, the reef offers plentiful soft corals and many interesting fish species.

10. **Sandy Deep**: Maximum depth 18 m. Experience level – Beginner. At the southern edge of the horseshoe-shaped plateau that surrounds Sandy Island this site is the most relaxing of the three Sandy Island sites with a moderate depth range in spite of its name.

11. **Sandy Shallow**: Maximum depth 20 m. Experience level – Beginner. Despite its name this reef drops down deeper than the other two Sandy Island dive sites. If your guide knows their way around properly then your safety stop can be spent in the shallow reef area that is only a few metres below the surface.

12. **M.V. Catheley H** (Wreck; Condition Moderate): Maximum depth 18 m. Experience level – Intermediate. There are many alternative spellings of this dive site that all phonetically sound the same. Whatever the spelling, this is a famous Anguillian wreck dive, being especially popular as a night dive because of its proximity to Sandy Ground. Good for larger species of fish.

13. **M.V. Lady Vi** (Wreck; Condition Good): Maximum depth 30 m. Experience level – Intermediate to Advanced. This wreck, relocated in February 2014 after a number of years 'lost at sea', is approximately 50 metres in length and still in relatively good condition. Home to many encrusting corals and sponges, with silvery fish shoals in abundance.

14. **M.V. Sarah** (Wreck; Condition Good): Maximum depth 25 m. Experience level – Intermediate to Advanced. At approximately 65 metres in length this is the largest wreck dive in Anguilla and also relatively intact - being upright and only moderately damaged. Impressive swim-throughs with cargo bays that take you deep into the heart of the vessel. One of the top dives in Anguilla.

15. **M.V. Meppel** (Wreck; Condition Good): Maximum depth 25 m. Experience level – Intermediate to Advanced. The Meppel, approximately 50 metres in length was relocated in 2010 after a number of years 'lost at sea'. As with the other wrecks in this area many shoaling pelagic species of fish can often be seen here together with numerous encrusting corals and sponges.

16. **M.V. Commerce** (Wreck; Condition Poor): Maximum depth 25 m. Experience level – Intermediate. This wreck is approximately 50 metres in length and still partially intact although recent storms have begun to take their toll. There is still plenty to explore and always a myriad of fish to appreciate.

17. **Hole-in-the-Wall**: Maximum depth 20 m. Experience level – Beginner. A relaxing start to this site takes you from the shallow mooring over a bed of soft corals on a gradually sloping incline. After a short distance the incline increases and suddenly drops impressively, offering some breath-taking scenery. The name of the site comes from a small but beautiful cave that is the perfect place to end your dive.

18. **Face-in-the-Wall**: Maximum depth 22 m. Experience level – Beginner. A shallow plateau at this site is a popular location to look for sleeping Nurse Sharks, with soft corals predominating towards the drop off a short distance away. One of the more popular dive sites along this part of the coast, it gets its name from the creepy face in the cliffs nearby. This site is also sometimes referred to as Captain Turtle.

19. **Shoal Bay**: In and around the Shoal Bay East area there were numerous once popular dive sites, although today they are seldom visited by dive operators due to travel distances and habitat degradation. These sites were known as Frenchmans East, Crystal Reef, Shoal Bay Reef, Lobster Reef, Sea Fan and Angel. Local dive operators know the location of all these sites and can arrange trips if specially requested. Many of the reef areas in and around Shoal Bay East are still well worth exploring.

20. **The Steps**: Maximum depth 30 m. Experience level – Intermediate. This is one of the top dive sites in Anguilla, and is located close to Little Scrub Island. It follows a system of ledges and huge boulders that form 'steps' taking you down to the deepest point where sharks and big pelagic fish are commonly sighted. Not for the faint hearted!

21. **Rams Head/Rock and Roll**: Maximum depth 30 m. Experience level – Advanced. The sites secondary name comes from the rough seas that are usually present off the point at Scrub Island, but it offers impressive underwater formations, some small caves, and the chance to see many large pelagic fish species. Also sometimes referred to as North Scrub.

Diving the M.V. Ida Maria (No. 8 previous page)

Diving the M.V. Oosterdiep (No. 6 previous page)

Not on Map:

22. **West Cay Dog Island**: Maximum depth 30 m. Experience level – Advanced. This site boasts one of the most impressive wall dives in Anguilla, although its depth and potentially strong currents can make it challenging. Usually undertaken as a drift dive, many species of large pelagic fish and other schooling species are present. One of the top dives in Anguilla.

23. **Devils Wall**: Maximum depth 30 m. Experience level – Advanced. Another impressive wall dive close to Dog Islands West Cay that offers wonderful opportunities to witness impressive underwater topography, sharks and other large pelagic fish. Not for the faint hearted as strong currents and rough seas can make it very challenging.

24. **Trigger Alley**: Maximum depth 20 m. Experience level – Intermediate. Close to East Cay this site boasts an alley-like gully that is home to many Black Durgons (page 190) who swim around in the water column. The sea here is rough much of the time, turning what could have been a beginner's dive into more of an intermediate challenge.

Disused Sites:

Over the years a number of sites that were once popular are now seldom visited by local dive operators. The main reason for this is habitat degradation offsetting the desire to travel the distances required. Aside from those sites at Shoal Bay East, this appears to be especially the case for five sites once located along the outer reef: Prickly Pear, Groupers Bowl, The Coliseum, Eel Garden, and Sand Canyon. Dougies Corner close to Sandy Island is worth a visit but its precise location is not widely known. Rita's Wall refers to the stretch of coastal reef towards Shoal Bay East that was subsequently split into separate sites. The Garden is a reef site to the northeast of the wreck of the M.V. Commerce. In terms of unconfirmed wrecks, reported sightings include: small yachts and similar close to Scrub Island; a car transporter off Rendezvous Bay in the St. Martin Channel; and a highly degraded plane close to Prickly Pear West. Any details relating to these or any other wrecks should be reported to the Department of Fisheries and Marine Resources. Many of the wrecks used as dive sites today were sunk deliberately to serve as artificial reefs after being damaged by past hurricanes. One of these wrecks, the Marva W, slipped into deep water off Meads Bay when being sunk. It was finally relocated in 2016, and found to be in excellent condition. Its 60 m depth however means it can only potentially be used as a technical dive site for experienced deep diver training.

1.9 Good Snorkelling Areas

A.   **Maundays Bay**: A relaxing snorkelling location and perfect for beginners as it follows the rocks in front of Pimms restaurant and is usually very calm. Home to many species of fish and other creatures. Continue along the rocks and the depth increases while the fish become more numerous and interesting. Don't expect a coral reef, but it is certainly worth investigating, with small caves and overhangs along the way.

B.   **Sandy Island**: After taking a water taxi from Sandy Ground to Sandy Island you will arrive on an archetypal desert island with a private bar and restaurant, plenty of sun, and of course lovely snorkelling. The best spots are to the north of the island, but watch out for sea urchins spiking your feet and delicate coral growths. It is best to enter the water to the west of the island and then make the short swim around to the north where beautiful corals and many shoaling fish can be seen. Look out for large colonies of finger corals (page 3).

C.   **Prickly Pear**: Arrange a day trip to the cays from Sandy Ground and visit this island paradise. There is a private bar and restaurant, deceptively strong sun and lovely snorkelling. To find the best spot it is best to avoid those areas where visitors are frequently seen snorkelling and walk along the beach towards its deserted eastern end. About half way along put on your mask, snorkel & fins and explore. There are some impressive Elk Horn colonies and a good diversity of fish to enjoy. At the right time of year this area is also well known for harmless baby reef sharks.

D.   **Little Bay**: Take a water taxi from Crocus Bay or another local charter to reach this beautiful and secluded spot. Once snorkelling do not expect to see a coral reef, instead this is a sandy bay with scattered rocks - but it does have a high diversity of fish species and is generally very enjoyable to explore. Don't forget to look up while snorkelling as there are impressive overhead rock formations to appreciate. This area is especially good for turtle spotting in early mornings or late afternoon. In 2019 a 100 m long snorkelling trail was set up in Little Bay by the Anguilla National Trust, in partnership with the Department of Fisheries and Marine Resources.

E.   **Forest Bay**: This is a little known snorkelling spot but worth a visit nonetheless. It is accessed by a small beach next to a fishing pier, and once you make it over the shallow seagrass the area gives way to a sandy bottom with impressive underwater coral heads. Care should be taken due to the occasional fishing boat, but these can be avoided by heading west once the water gets a bit deeper. In this area the sand gives way to a once impressive reef system split in two by a deep sand channel. The eastern edge of this channel offers dramatic underwater scenery, although this snorkel should only be undertaken by confident swimmers.

F.   **Shoal Bay East**: Just off the point of this beautiful beach is probably the best snorkelling on Anguilla, although it is worth asking local establishments as to precisely where to go. Access to the area is by way of small channels which can be hard to locate, but the place you should be aiming for is the outside of the closest reef to the coast, not the more distant Madeariman Reef. The reef here is shallow and drops down into a sand channel, but the slope is gentle enough to allow a snorkeller to witness all the action. There is usually a vast array of fish, some beautiful coral formations, and a relatively good chance of seeing a foraging Hawksbill turtle. Sadly, as with all of the reefs around Anguilla, the overall health of this area is not what it was a few decades ago.

View of Sandy Island Marine Park in 2006

View of Little Bay Marine Park in 2006

Chapter 2 - Plants and Algae

Y-branched algae (*Dictyota* sp.) - page 17

This chapter details the small selection of marine flowering plants that are found in Anguillian waters combined with the most common algal species. Algae taxonomy and their place, or lack there of, in the Kingdom Plantae is still being debated. One of the main sticking points in this debate is whether or not each of the algal groups arose independently from separate non-photosynthetic ancestors, as this ability - to convert sunlight into chemical energy - is the common link between the groups. This photosynthetic quality was achieved when cyanobacteria (page 28) were engulfed aeons ago by ancestral organisms that went on to incorporate them into their own cell design. These engulfed cyanobacteria became chloroplasts, the organelles in modern plant/algae cells that are responsible for photosynthesis. Cyanobacteria, also known as blue-green algae, are still present today and are an independent and important group in their own right (page 28).

Many species of marine algae are free-floating single-celled phytoplankton. These phytoplankton form the base of the oceanic food chain but are microscopic and so not included in this chapter. Furthermore, as coral reef ecosystems are historically oligotrophic they lack the nutrients to support phytoplankton in large enough numbers for them to be noticeable. If present in sufficient numbers they can give sea water a green tinge, a phenomenon that appears to be becoming increasingly common in the Caribbean. This suggests the balance of low nutrient levels may be in the process of shifting which will have huge consequences for coral reefs, a topic that will be discussed further in Chapter 6 (page 220-230).

This chapter has been organised without entering into the debate or complexity of what constitutes a plant, although it has been laid out with the true plants occupying the first section and subsequent sections featuring phyla that get progressively less plant-like from a morphological standpoint. Within each section only the most commonly occurring species have been listed, and it should thus be appreciated that floral diversity is potentially much broader than might first be supposed. Also, as species are sometimes very difficult to identify, identification may only be down to Genus level in many cases, with a selection of photographs providing further examples. The section headings that divide this chapter are listed below:

2.1 Flowering Plants

2.2 Brown Algae

2.3 Green Algae

2.4 Red Algae

2.5 Blue-green Algae (Cyanobacteria)

It should be noted that the term division is usually used in plant taxonomy instead of the term Phylum, although in some classification systems it is slowly being adopted. However, for the purpose of this book the following plant section has been labelled as a division whereas the subsequent algal sections have been classed as phyla. For those interested in the complex current debates on plant taxonomy a more detailed text on the subject should be sought. Please refer to Chapter 6 for suggested reading (page 228).

This chapter does not cover **turf algae** in any detail (see photograph below), as the group comprises more than 100 species in the western Atlantic and identification is extremely difficult, usually requiring microscopic examination by an expert. Turf algae are often referred to as microalgae because of their small size, whereas the larger algal types covered in this chapter are generically known as macroalgae. Turf algae are usually filamentous, attaining a height between 1 to 10 mm and forming a lawn-like covering over rocks. Only a few species are able to persist or remain abundant throughout the year, with most others having a generally high turnover, although as a group their overall cover on rock remains relatively stable year round. Turf algae are an important and constant food source for grazers such as sea urchins (page 98-102), mainly because the algae can recover rapidly after being partially consumed. Turf algae are capable of trapping ambient sediment and can kill corals by gradual encroachment (Chapter 6 - page 220-230), so it is important to maintain the grazer/turf algae balance. Newly uncovered rock or reef remains bare for extremely short periods as it is rapidly covered by turf algae, sediment, reef cement, crustose coralline algae, or a mixture of these, before potentially being colonised by larger organisms.

Turf algae often forms a dense sediment-rich mat covering rock or dead coral

2.1 Division – Angiospermae: Flowering Plants

The angiosperms, also sometimes referred to as the Magnoliophyta, is a taxonomical group whose classification is under revision. It is currently classed as an unranked division within the spermatophytes (seed-bearing plants), who themselves form a traditional division within the plant Kingdom. There are only a few species of marine flowering plants, all of which have true roots, stems, leaves, and flowers like the more familiar terrestrial flowering plants. Those most commonly observed by divers and snorkellers are the seagrasses that can cover huge areas of sand forming important habitat for herbivores (e.g. Green Turtle – page 207) and juvenile reef fish (e.g. Yellowtail Snapper – page 132).

Common Name – **Turtle Grass**

Scientific Name – *Thalassia testudinum*

Description – Green, flat, erect leaves/blades of equal width along their length. Has an extensive root system that gives effective anchoring and helps stabilise sand/mud bottoms. Flowers are pinkish green-white, but uncommonly sighted (see inset image). Leaves/blades are often covered with sediment or epiphytes. This is the most common species of flowering plant in Anguillian waters, forming many large dense beds in areas such as Little Bay, Rendezvous Bay and Road Bay. Is most common in depths less than 10 m. Blades can be over 25 cm in length and 1.5 cm in width.

Common Name – **Manatee Grass**

Scientific Name – *Syringodium filiforme*

Description – Green, thin, cylindrical leaves/blades that are generally erect. Has an extensive network of runners and often occur mixed in with Turtle Grass (previous) as pictured to the right, although may also form exclusive beds of their own. Common, growing on sandy bottoms, with blades sometimes longer than 25 cm.

Common Name – **Midrib Seagrass**

Scientific Name – *Halophila baillonis*

Description – Green, oval, erect leaves with an obvious midrib line. Has an extensive root system and prefers deeper sandy/muddy habitats. Identification of pictured specimen is tentative as *H. stipulacea*, an invasive species of seagrass from the Indian Ocean, is almost identical in appearance. This species is able to colonise sandy areas unsuitable for *H. baillonis*, so its spread is positive in terms of sand stabilisation – especially considering potential increases in storm frequency due to climate change. Increased fish biomass on colonised areas, and more foraging grounds for herbivorous species (including sea turtles) are also being reported as positive effects. Prior to 2010, with the exception of some deeper areas in the St. Martin Channel and close to the wreck of the M.V. Oosterdiep, *Halophila* sp. was only found in small quantities. In 2015, seagrass monitoring conducted by the Department of Fisheries and Marine Resources recorded a considerable increase in abundance within most areas studied.

2.2 Phylum – Phaeophyta: Brown Algae

Brown algae range in colour from pale tan to greenish brown to dark brown, although some species may even be iridescent blue. The brown colouration is the result of a pigment known as fucoxanthin. The best known members of this group are the classic seaweeds and kelps that form dominant littoral zone communities in many temperate areas, although several species of brown algae are also common inhabitants of tropical coral reefs.

Common Name – **Sargassum Weed**

Scientific Name – *Sargassum* sp.

Description – The most common species, *S. fluitans* (Sargassum Seaweed) and *S. natans* (Sargasso Weed), are tan to green in colour and float freely on the surface of the water using small gas-filled floats for buoyancy. Other species fix themselves to the sea floor (Sargassum Algae), as in the pictured example. Common, with individual plants up to 45 cm in size. Beginning in 2014 huge quantities of the free-floating species began washing up in the Caribbean, reportedly as a result of increased nutrient levels arriving into the Caribbean from the Amazon and Orinoco river plumes. Levels are normally much lower and originate from the Gulf of Mexico. The inundation lead to smelly rotting piles of biomass on otherwise pristine tourist beaches, and floating mats in the water clogged boat engines. By the end of 2015 inundation levels appeared to be dropping but picked up again the following year (page 221).

Common Name – **Y-Branched Alga**

Scientific Name – *Dictyota* sp.

Description – This Genus contains several species that are difficult to distinguish, some of which have yet to be described by scientists. A general identification feature for the group is the presence of branches that are forked at the end, sometimes more than once. Colours vary from tan to green to iridescent blue. Some species have quite wide branches, others thin, and some that are serrated. This Genus is probably the most common algae growing on Anguillian reefs, in many parts visibly smothering coral – a subject that is discussed further in Chapter 6 (page 220-230). Locally abundant, but does not usually exceed 10 cm in height.

Common Name – **Leafy Flat-Blade Alga**

Scientific Name – *Stypopodium zonale*

Description – Small bush-like plant with relatively long, flat, square-ended blades that are irregularly branched and patterned with thin dark bands. Comes in a variety of colours but is usually shades of green and similar in appearance to Leafy Rolled-Blade Alga (next). *Stypopodium zonale* is distinguished by long flaccid blades with dark-coloured banding. Attaches to rock and is locally common. Blade length can be over 25 cm.

Common Name – **Leafy Rolled-Blade Alga**

Scientific Name – *Padina boergesenii*

Description – Forms clumps of low-lying leafy blades, often slightly semi-circular in shape. Blades are irregularly branched and usually green in colour with thin light bands. Is similar to Leafy Flat-Blade Alga (previous), although distinguished by firmer, shorter blades with light-coloured banding. Other species in this Genus look similar to the pictured specimen so identification is not certain. Attaches to rock and can be locally common. Blade length is not usually longer than 7 cm.

Common Name – **White Scroll Alga**

Scientific Name – *Padina jamaicensis*

Description – Forms clumps of low-lying leafy blades, often slightly semi-circular in shape. Blades are irregularly branched and white in colour with light banding. Other species in this Genus look similar to the pictured specimen so identification is not certain. Attaches to rocky surfaces and can be locally abundant. Blade length is not usually longer than 5 cm.

Common Name – **Encrusting Fan-Leaf Alga**

Scientific Name – *Lobophora variegata*

Description – Has thin, fan-shaped blades that have an encrusting appearance, often overlapping in a scale-like fashion. Usually green in colour although can be brownish. Usually favours shady areas where other algae do not readily colonise due to lack of light. Blade surface is often covered with sediment and/or epiphytes. Abundant, growing on most reef habitats and especially common in parts of Shoal Bay East. Usually do not exceed 10 cm in diameter.

Common Name – **Saucer Leaf Alga**

Scientific Name – *Turbinaria* sp.

Description – Thick, triangular, cone-shaped blades with saucer-like tips attached centrally, and in clumps, to an erect central column. Is usually greenish-tan in colour and is occasional on shallow areas with moderate to strong water movement. Saucers are not more than one or two centimetres in diameter.

2.3 Phylum – Chlorophyta: Green Algae

Green algae were historically the most commonly observed algae on tropical reefs, although in recent years brown algae (previous) seem to have taken their crown, at least in Anguillian waters. Many green algae are calcareous (e.g. *Halimeda* sp.), having an internal calcium carbonate structure around which the algae grows. *Halimeda* sp. adds significant amounts of calcium carbonate to the reefs as this structure remains intact after their death, and can eventually wash up along the coast and contribute significantly to beach material. Some beaches are in fact almost exclusively made up of their remains (see for example parts of Shoal Bay East), and are as important for sand formation as parrotfish excreta (page 154). Chlorophyll is the pigment responsible for the colour of this algal group, which varies from bright to dark green.

Common Name – **Watercress Alga**

Scientific Name – *Halimeda opuntia*

Description – Thick, dense, low-lying clumps of rounded leaf-like segments, dark to bright green in colour. Grows in shallow depressions, gaps, and other vaguely protected areas. The dead calcified segments of this and other *Halimeda* sp. are a major contributor of calcium carbonate to the reefs and sand. Common, with segments no wider than 1 cm.

Common Name – **Stalked Lettuce Leaf Alga**

Scientific Name – *Halimeda tuna*

Description – Small branched clumps of thin rounded segments arise from a single stalk of segments, usually bright green in colour. Grows on sand, amongst seagrasses, or on rubble and reefs. The dead calcified segments of this and other *Halimeda* sp. are a major contributor of calcium carbonate to the reefs and sand. Locally abundant but generally common. Usually around 10 cm tall with segments not much wider than 1 cm.

Common Name – **Three Finger Leaf Alga**

Scientific Name – *Halimeda incrassata*

Description – Small branched clumps of thin segments with three distinct lobes on each arising from a branched stalk of segments. Is dark to bright green in colour and grows on sand and in amongst seagrasses, although is sometimes found on rubble and reefs. The dead calcified segments of this and other *Halimeda* sp. are a major contributor of calcium carbonate to the reefs and sand. Locally common but generally occasional. Usually around 10 cm tall with segments not much wider than 1 cm.

Common Name – **Green Jointed-Stalk Alga**

Scientific Name – *Halimeda monile*

Description – Small, upright cylindrical but slightly flattened segments with occasional branching, dark to bright green in colour. Prefers shallow sandy areas, especially seagrass beds, although also often grows on reefs. The dead calcified segments of this and other *Halimeda* sp. are a major contributor of calcium carbonate to the reefs and sand. Locally common but generally occasional. Usually not much taller than 10 cm.

Notes on other *Halimeda* species: Although other *Halimeda* species are present in Anguillian waters, for example **Large Leaf Watercress Alga** (*H. discoidea* - below left), at times identification can be difficult due to the variety of growth forms present (below middle) so a full list has yet to be compiled. *Halimeda* sp. can also form expansive beds, as in the pictured example (below right) seen in Sandy Hill Bay.

H. discoidea (probable ID)

Unconfirmed species, although very similar to *H. tuna*

Halimeda bed at Sandy Hill Bay

Common Name – **Bristle Brush Alga**

Scientific Name – *Penicillus* sp.

Description – Paintbrush-like in appearance with a short stalk anchored in the sand. Darkish green with tightly packed bristle-like filaments. Identification down to species level is problematic, sometimes requiring laboratory examination, although *Penicillus pyriformis* (usually with a flat top), *Penicillus capitatus* (thin stalk rounded top – probably most common species in Anguilla) and *Penicillus dumetosus* (thick stalk, large rounded top) are the most common members of this Genus found in the Caribbean. Common to occasional on sandy bottoms and seagrass beds, not usually growing more than 10 cm in height.

Penicillus capitatus (probable ID)   *Penicillus dumetosus* (probable ID)   *Penicillus pyriformis* (probable ID)

Common Name – **Green Feather Alga**

Scientific Name – *Caulerpa sertularioides*

Description – Rigid feather-like structures grow upright from long, cylindrical runners. Bright green in colour, growing predominantly among seagrasses and on shallow sandy areas. Occasional, with 'feathers' usually less than 10 cm in height.

Common Name – **Flat Green Feather Alga**

Scientific Name – *Caulerpa mexicana*

Description – Rigid, flat feather-like structures grow from long, cylindrical runners. Bright green in colour, growing predominantly among seagrasses. This species is especially noticeable in areas of Road Bay. Locally occasional, with blades not much longer than 3 to 4 cm.

Common Name – **Oval-Blade Alga**

Scientific Name – *Caulerpa prolifera*

Description – Flat, twisted oval blades grow from cylindrical stalks. Dark green in colour and easily mistaken for the seagrass amongst which they grow. This species is especially noticeable in areas of Road Bay. Locally common, with blades not much longer than 3 to 4 cm.

Common Name – **Green Grape Alga**

Scientific Name – *Caulerpa racemosa*

Description – Clusters of grape-like spherical structures attach to a long cylindrical runner via thin branches. Can form what looks like one big cluster when densely clumped, and is usually bright to medium green in colour. Prefers hard rocky bottoms where it often lives in crevices, although it does seem to prefer some water movement. Occasional, with 'grapes' not much bigger than 0.5 cm, and clusters not usually more than 10 cm in height.

Notes on other *Caulerpa* species: This Genus probably contains the greatest number of species of any algae found in Anguillian waters. Others that are likely to be seen include: **Cactus Tree Alga** (*Caulerpa cupressoides*) - below left; the thin paintbrush-like forms of *Caulerpa lanuginosa* (not pictured) that often grows cryptically within seagrass beds; and the delicate circular frills of *Caulerpa verticillata* (below middle). Also pictured here is an unidentified species of *Caulerpa* that was seen abundantly in Road Bay during 2009 - below right. The serrated edges suggest it may be a closely related upright variety of **Saw Blade Alga** (*Caulerpa serrulata*) - not pictured.

Caulerpa cupressoides

Caulerpa verticillata

Caulerpa sp.

Common Name – **Fuzzy Finger Alga**

Scientific Name – *Dasycladus vermicularis*

Description – Cylindrical branches with thin fuzzy filaments covering the tips. Occasional, although may be abundant in areas of elevated water nutrients. This species will grow on almost any surface, with each branch not exceeding a few centimetres in length. Other unconfirmed species of similar appearance from a variety of Genera may also be present in Anguillian waters.

Common Name – **Sea Pearl**

Scientific Name – *Ventricaria ventricosa*

Description – Shiny dark green spheres often covered with patches of Lavender Crust Alga (page 27). These spheres are in fact a single cell, and are one of the largest single celled life forms in either the plant or animal Kingdom. They attach to hard surfaces, often in crevices, via fine hair-like structures, and although usually solitary can grow in groups. Transparent individuals are often seen floating free after death. Common, although are often overlooked when in crevices. Can grow to over 5 cm in diameter.

Common Name – **Saucer/Paddle Blade Alga**

Scientific Name – *Avrainvillea* sp.

Description – A fan-shaped blade radiates out from a thick stalk that is anchored in the sand, which is in turn connected to long slender runners. Blade surfaces are smooth (distinguishing them from *Udotea* sp.). Of the two species *Avrainvillea asarifolia* has a rounded blade underside, and the smaller *Avrainvillea longicaulis* generally does not, although positive identification between these and other species within the Genus requires microscopic examination. Occasional, not usually growing more than 10 cm tall.

Photo Credit Stuart Wynne

Common Name – **Mermaid's Fan**

Scientific Name – *Udotea* sp.

Broad fan-shaped blades similar in appearance to *Avrainvillea,* although surfaces are rough due to heavier calcification. These surface features may be straight lines radiating up the blade, and/or concentric circular ones across it. There are a number of species in this Genus that are similar in appearance and/or take on a variety of growth forms: Large wavy-edged blades; flat single blades; bisected multi-blades; and a cup-shaped species. Blade arrangements originate from a single stem. Prefers sandy bottoms and seagrass beds, and usually grows in groups. Locally common to occasional, usually not growing more than 15 cm in height, although this varies among species.

Photo Credit Stuart Wynne

Common Name – **Mermaid's Wine Glass**

Scientific Name – *Acetabularia* sp.

Description – Small, saucer-shaped, almost mushroom-like caps sit on top of short thin stems with radiating ridges that give the appearance of a fluted wine glass. Of the two known species *Acetabularia calyculus* (pictured) is green and gregarious, whereas *Acetabularia crenulata* (not pictured) is white and solitary. Both species grow in sheltered habitats where they can be locally common, for example at Maundays Bay. *A. crenulata*, the larger of the two species, may reach 5 cm in height.

Photo Credit Stuart Wynne

Common Name – **Green Filamentous Alga**

Scientific Name – *Chaetomorpha linum*

Description – Forms distinctive entangled matted structures that are green to yellow-green in colour and made up of hair-like cylindrical filaments. These filaments are each a single row of cells, and often tangle up into balls as they roll around freely on the sea floor. They have a coarse steel wool texture especially when washed up, and are often associated with high nutrient areas. Large blooms of this species sometimes wash up on the northern end of Sandy Ground, probably caused by nutrient leeching from the nearby salt pond. The species is uncommon in Anguilla. Balls vary in size but are usually not much larger than 15 cm. Identification is probable, as there are a number of similarly appearing species that require microscopic examination to differentiate.

2.4 Phylum – Rhodophyta: Red Algae

Red algae are often abundant on tropical reefs but go unnoticed because of their sometimes small size, indistinct growth forms and dull colours. They will often be the first species to colonise a mooring rope, forming a thin, filamentous, mat-like cover over the rope that can irritate bare skin. Other species are encrusting, and form thin calcified patches on the surface of rock that would never be recognised as algae without microscopic examination. These encrusting species are very important for reef ecology as they cement broken coral fragments together, thus strengthening the reef structure. This is especially important after hurricanes. Other calcareous varieties form twig-like thickets. The reddish colouration is the result of a pigment known as phycoerythrin.

Common Name – **Thicket/Twig Alga**

Scientific Name – *Amphiroa* sp., *Jania* sp., & *Galaxaura* sp.

Description – Members of these three Genera look similar, forming calcareous branches of varying thicknesses. *Galaxaura* sp. usually forms the thickest branches, and *Jania* sp. the thinnest. Of all Genera *Amphiroa* sp. is the most common (pictured), although identification to species level is very difficult and requires microscopic examination. Thickets are not usually larger than 10 cm in height, again with *Jania* sp. being the smallest. Occasional, although very small individuals are probably common but are usually overlooked as they frequently grow in crevices.

Common Name – **Limu Weed**

Scientific Name – *Asparagopsis taxiformis*

Description – Greenish to red-grey upright branches comprised of such fine filamentous subdivisions that each 'frond' takes on a fluffy appearance. This species is distinctive and, interestingly, wasn't reported in Anguilla until 2008 when it was sighted in one particular location at Shoal Bay East where it is now considered common. Has yet to be confirmed in other areas. The common name comes from the Hawaiian word for seaweed where this species' strong penetrating flavour is used in cooking, although it has to be soaked overnight to reduce its bitter iodine taste. Maximum frond length is c.20 cm.

Common Name – **Stub-Branched Alga**

Scientific Name – *Laurencia* sp.

Description – Usually red to tan in colour, this Genus contains a number of similar-looking species that are difficult to distinguish underwater. Generally species have smooth cylindrical branches with short stubby branchlets and rounded tips. Occasional, not usually growing more than 10 cm in height.

Common Name – **Spiky Red Alga**

Scientific Name – *Bryothamnion triquetrum*

Description – This coarse-looking red algae has many stiff short branches that are forked at the tips giving a very spiked appearance. It is usually found on shallow rocks exposed to wave action and so is unlikely to be seen by snorkellers but is of interest because it is often used to bait fish traps when targeting herbivorous species. It may also be seen growing on mooring lines and can cause slight skin irritation if it scratches the surface.

Common Name – **Crustose Coralline Algae**

Scientific Name – Various

Description – Forms thin, brittle, calcified encrustations that often have a leaf-like appearance, sometimes overlapping, but may also be flat. Colourations vary greatly both within and between species, and growth forms are so individually variable that microscopic examination is required to identify to which Genus a specimen belongs. Encrusts rock and seems to prefer shaded regions, often in cracks and under ledges. Crustose coralline algae play an important role in the coral reef community by contributing calcium carbonate to reef structure and studies have suggested that they also encourage settlement of juvenile corals to the rock surface. Can be locally abundant, although in most reef areas is considered common to occasional.

Common Name – **Reef Cement**

Scientific Name – *Porolithon pachydermum*

Description – Forms thin, highly calcified encrustations, and takes on the surface texture of the rock on which it is growing. In fact, although to the untrained eye rocks often look devoid of growth, this is extremely rare, and it is more likely that the seemingly bare area is in fact covered with this Reef Cement and/or turf algae (page 14). Its common name relates to the fact that layers of this species are laid down over rocks and effectively cement them together over a number of years, a habit that is extremely important in stabilising fractured or fragmented reef structures. Abundant.

Common Name – **Lavender Crust Algae**

Scientific Name – Various

Description – Similar in many respects to Reef Cement (previous), these species belong to a variety of Genera that form small patches of highly calcified encrustations usually lavender in colour. These species are considered abundant and can be seen growing on a variety of other organisms including Turtle Grass, Sea Pearls and Blade Algae.

Notes on other red algal species: The species listed above represent only a selection of red algae species that live in Anguillian waters and as with other algal groups, identification can at times be problematic without examining specimens in a laboratory. Below are pictured three such examples.

Unconfirmed red algae species

Unconfirmed red algae species

Unconfirmed Crustose Coralline Algae species

2.5 Phylum – Cyanophyta: Blue-green Algae

The blue-green algae are somewhat out of place in this chapter as they are not plants at all, and not even algae as their common name suggests. It is more accurate to call them by their modern name, cyanobacteria, as this correctly identifies them as a type of bacteria. The reason they have been included in this chapter is because they photosynthesize, grow in observably plant-like forms, and are extremely significant organisms in their own right.

Cyanobacteria have been present on the earth for almost 3 billion years, making them the most ancient form of life still alive today. Furthermore, their activities on this early Earth probably led to the conversion of our atmosphere into one that is rich in oxygen, making life as we know it possible. As if this achievement were not great enough, plants and algae that can photosynthesize possibly gained the ability through the endosymbiosis of cyanobacteria, a process where an organism is absorbed onto another but continues to function and eventually becomes an integral part of it. All in all, without cyanobacteria, the world would be a very different place.

Having said this, cyanobacteria in the tropics can cause events known as toxic tides which produce bad odours and can kill local fish populations. Although this has yet to be documented in Anguilla, an increasing abundance of cyanobacteria has been noticed over the last couple of years and concerns exist that such toxic conditions will one day become common occurrences. This will not only affect local fishing livelihoods but also make certain parts of the island much less attractive to the essential tourist trade. As cyanobacteria grow more rapidly in nutrient-rich conditions their presence can be an indication of eutrophication (Chapter 6 - page 220-230), and hence eutrophication can be a contributing factor to these toxic tides.

The species of cyanobacteria present in Anguillian waters are not possible to identify without microscopic examination by an expert. Although some species are unicellular, the ones of interest here are multicellular, usually growing in long chains. They normally take the form of fuzzy, mat-like structures and are usually brown in colour. Growth forms vary from a thin mat covering patches of seabed to quite large clumps that can overgrow coral species, smothering the coral and killing part of the colony. Cyanobacteria are also responsible for many coral diseases (page 69-70), and thus can result in not just the death of a complete colony but many colonies in a particular region.

Different cyanobacteria growths: fuzzy ball (top left); mat-like growths (top right); and attached to soft corals (bottom) - Photo Credit Stuart Wynne

Chapter 3 - Invertebrates

Invertebrates are an informal, generic group of the animal Kingdom whose unifying characteristic is that they lack a back bone. The back bone is made up of smaller connected bones known as the vertebrae, and although some invertebrates do have a vestige of this supporting structure during some stages in their lifecycle, the majority do not. Instead, most have soft bodies that may or may not be protected by an external protective skeleton. Of course, nature's diversity means the boundary between these two major groups undergoes some confusion, as can be illustrated by tunicates (page 104-105) and jawless fish (not in this book), who belong to the Phylum Chordata together with all the vertebrates. Tunicates look very much like primitive creatures, although as larvae they have a semblance of a primitive backbone. The jawless fish (i.e. Lampreys & hagfish), although very fish-like, lack a backbone instead having what is thought to be its precursor, the notochord.

There is great variety among the invertebrates, from the sessile sponges (page 30-38) to the ecosystem building hard corals (page 51-64) and the armoured crustaceans (page 75-82). For ease of reference these have been arranged taxonomically, beginning with the simplest phyla, and progressing in complexity. This progression is somewhat subjective as there are many ways to categorise the complexity of a life form, although in terms of its use here it is meant in the more anatomical sense as set down by Humann & Deloach (1992). For example, bryozoans (page 84) look like relatively simple creatures related to coral, they are in fact far more complex and actually more barnacle-like in many ways, hence their position after the crustaceans. Similarly, although flatworms (page 71) are often considered very simple creatures, their independent and mobile nature has placed them above the Cnidarians (page 39-70). Below is a contents list for the main section headings.

3.1 Phylum – Porifera: Sponges

3.2 Phylum – Cnidaria: Cnidarians (including hydroids, jellyfish, anemones, zoanthids and corals)

3.3 Phylum – Ctenophora: Comb Jellies

3.4 Phylum – Platyhelminthes: Flatworms

3.5 Phylum – Annelida: Segmented Worms

3.6 Phylum – Arthropoda: Arthropods (Including crustaceans)

3.7 Phylum – Ectoprocta: Bryozoans

3.8 Phylum – Mollusca: Molluscs (Including gastropods, bivalves, squid and octopuses)

3.9 Phylum – Echinodermata: Echinoderms (Including sea urchins, sea stars, brittle stars and sea cucumbers)

3.10 Phylum – Chordata: Chordates (Tunicates)

Identification note: In contrast to the other chapters in this book, maximum sizes have been omitted from certain sections because of the group's nature. For example, sponges can essentially keep growing until something either eats or breaks them, and corals are tiny creatures that grow in colonies of uncertain sizes. Having said this, in these sections generic comments on size may be made if it is a useful key to distinguishing between two visually similar species.

3.1 Phylum – Porifera: Sponges

Sponges are considered to be one of the most primitive groups of multicellular animals. This is arguably untrue because sponges represent what is in effect an evolutionary 'dead-end', and is thus considered by some as a perfect design as no significant modifications have been made for untold millennia. Sponges are sessile with the ability to grow on a variety of surfaces where they filter feed by drawing water into numerous small holes on their surface called in-current pores or openings (ostia). This water is moved through the sponge by currents created by beating, whip-like extensions on the individual cells called flagella. As the water passes through the sponge food and oxygen are filtered out and the water leaves the internal cavity via one of the animal's larger ex-current openings (oscula). These are usually further towards the top of the sponge than the in-current openings to aid natural circulation. Sponges lack muscle, nerves, internal organs or true tissue layers, and their individual cells display a considerable degree of independence. This independence is at such a level that if a sponge is passed through a cheese cloth back into sea water and reduced to individual cells, the cells can cluster together and begin building a new sponge structure – a process called cloning.

Sponges come in many shapes and sizes but can be recognised as such by their ex-current openings, although they should not be confused with the tunicates (page 104-105), who, unlike sponges, react when disturbed by closing their openings. However, identification down to species is sometimes impossible underwater, and can only happen if a sample is analysed in a laboratory. Here, under a microscope, the skeletal spicules can be examined allowing the species to be determined. These spicules are made of calcium carbonate or silica, and are often quite brittle. Some sponges however have a softer skeleton made of spongin, and are the more familiar types used to bathe with. Identification underwater is further hindered as a single species can take on a variety of shapes and colours which are influenced by factors such as age, water movement, water chemistry, light conditions or the presence of algae living in symbiosis.

For this reason only the more common species that grow in relatively consistent body-patterns are described here, and they have been grouped together by morphology rather than scientific classification. Photographs of some unidentified sponges have been included at the end of this section to illustrate that the species present in Anguillian waters are very diverse, and as such can not be represented fully within this guide book.

What role do sponges play in coral reef ecosystems? Sponges look like inert creatures that barely fit into the animal Kingdom, instead seeming to belong to some strange group of plants, or even a Kingdom of their own. Until taxonomists decide otherwise, they remain classified as animals (see above) and play an important role in coral reef ecosystems, one that over recent years is being suggested to be more significant than ever before thought. It is known that they filter huge amounts of water through their structure, and play a potentially vital role in reducing turbidity, thus helping coral reef waters to remain clear. However, it is also believed that they play what might be considered a negative role in terms of the phase-shift between corals and algae that has taken place over the last few decades (Chapter 6 - page 220-230). Researchers suggest that a feedback loop exists between algae and sponges, where algae release dissolved oxygen that sponges consume. The sponges then in turn return carbon to the reef but also release nutrients that further enhance algae growth. To add to this, both algae and sponges compete for space with corals, and the cycling carbon and nutrients alter microbial activity that also has negative consequences for the coral reef ecosystem. So are sponges in actual fact bad news for coral reefs? Not necessarily, as their filter feeding nature might outweigh the overall feedback effect, but it is certainly a research topic that coral reef biologists need to understand in more detail.

Common Name – **Branching Tube Sponge**

Scientific Name – *Pseudoceratina crassa*

Description – Clusters of numerous tubes extend from the base, but in Anguillian waters they are usually quite low-lying, with a profusion of small bumps on their exterior, giving a stippled texture. Morphology can vary greatly, and colour ranges from yellow to purple to olive green. Thus, although common, this species is not easily identifiable.

Common Name – **Yellow Tube Sponge**

Scientific Name – *Aplysina fistularis*

Description – Clusters of numerous tubes extend from the base, and are usually yellow in colour, although they may have a green tint at times. Tubes are soft walled, and often have antler-like growths protruding from them. Common.

Common Name – **Branching Vase Sponge**

Scientific Name – *Callyspongia vaginalis*

Description – Clusters of tubes extend from the base, although can have solitary tubes. Tube walls are rigid and thin, often with a stippled surface. Can form fan-like shapes (not pictured) and are usually purple in colour although this often varies from brownish to greenish grey. This is one of the most common sponges in Anguillian waters, and is usually associated with the Sponge Zoanthid (*Parazoanthus parasiticus* – page 48) and also sometimes the Sponge Brittle Star (*Ophiothrix suensonii* – page 97) as shown here.

Common Name – **Azure Vase Sponge**

Scientific Name – *Callyspongia plicifera*

Description – Pinkish-purple to almost florescent blue vase-like sponge. Its exterior is elaborately coloured, and may live in association with the Sponge Brittle Star (*Ophiothrix suensonii* – page 97). Uncommon.

Common Name – **Pink Vase Sponge**

Scientific Name – *Niphates digitalis*

Description – Pinkish to greenish-grey in colour with a bowl-shaped appearance. The bowl edge is bordered with an almost transparent membrane, and the surface is very rough and pitted, usually covered with Sponge Zoanthids (*Parazoanthus parasiticus* – page 48). The Sponge Brittle Star (*Ophiothrix suensonii* – page 97) also often lives in association with it. Occasional.

Common Name – **Netted Barrel Sponge**

Scientific Name – *Verongula gigantea*

Description – Can form quite large barrel shapes with a distinctive raised net-like surface structure. Can vary in colour from yellowish-green to dark brown. The pictured specimen was sighted in the St. Martin Channel at a depth of c.30 m. Uncommon.

Common Name – **Giant Barrel Sponge**

Scientific Name – *Xestospongia muta*

Description – Usually a single, huge barrel-shaped sponge, although occasionally may form clusters. Has a very hard jagged surface and is dark brown in colour. Uncommon on most reef habitats, although can be locally common, for example around Anguillita Island.

Common Name – **Black Ball Sponge**

Scientific Name – *Ircinia strobilina*

Description – Ball-shaped with stippled surface along with clusters of ex-current pores. As its common name suggests it is generally black in colour although stipples may be paler. Common to occasional.

Common Name – **Orange Ball Sponge**

Scientific Name – *Cinachyra* sp.

Description – Orange, round and pitted with numerous ex-current openings. Usually has a covering of sediment or algae. Two species in this Genus look almost identical and so positive identification outside of a laboratory is not possible. Common on most reef environments.

Common Name – **Lumpy Overgrowing Sponge**

Scientific Name – *Holopsamma helwigi*

Description – Usually pinkish although colour can vary. Grows as a lumpy sometimes rope-like mass often using gorgonian stalks as a support, as in the pictured specimen sighted at Limestone Bay. Considered occasional on reefs that have an abundance of soft corals.

Common Name – **Lavender Rope Sponge**

Scientific Name – *Niphates erecta*

Description – Long and rope-like with scattered pores and an uneven pitted surface. Lavender to pink in colour and usually associated with the Sponge Zoanthid (*Parazoanthus parasiticus* – page 48) and also sometimes the Sponge Brittle Star (*Ophiothrix suensonii* – page 97). Occasional.

Common Name – **Brown Encrusting Octopus Sponge**

Scientific Name – *Ectyoplasia ferox*

Description – Reddish to dark brown with raised ex-current openings that have a light-coloured ring around their edges. Generally lumpy, this species can take on many growth forms as is common among sponges, and needs microscopic examination to confirm identification. For this reason identification of the pictured specimen is tentative. Common to occasional.

Common Name – **Row Pore Rope Sponge**

Scientific Name – *Aplysina cauliformis*

Description – Usually a twisted rope-like structure with ex-current openings in distinct rows, whose protruding lips are of lighter colour. Has a smooth surface and a highly variable colour, although is most commonly purple to red. Common on reefs, with smaller, usually dark purple individuals found inhabiting seagrass beds.

Common Name – **Scattered Pore Rope Sponge**

Scientific Name – *Aplysina fulva*

Description – Usually a twisted rope-like structure with ex-current openings not forming distinct rows and having lips that protrude. Has a smooth surface and a highly variable colour, although is most commonly yellow to tan. The protruding lips are often lighter in colour than the rest of the sponge. Occasional.

Common Name – **Green Finger Sponge**

Scientific Name – *Iotrochota birotulata*

Description – Shades of green with lumpy finger-like branches that may grow erect or form tangled masses. If squeezed will secrete a black dye. Occasional.

Common Name – **Erect Rope Sponge**

Scientific Name – *Amphimedon compressa*

Description – Erect and rope-like as its common name suggests, with a smooth but porous-looking surface that is a good key to identification. Usually red in colour but can vary towards purple. Ex-current openings do not have differently coloured lips. Occasional.

Common Name – **Brown Boring Sponge**

Scientific Name – *Cliona tenuis*

Description – Dark brown, encrusting relatively large areas of rock in shallower regions. Bores into rock under the ex-current openings, and after death leaves the rock with a series of holes. These holes accelerate erosion, and thus this species is ecologically very important, especially in terms of reef degradation (Chapter 6 - page 220-230). Common in shallow areas.

Common Name – **White Convoluted Sponge**

Scientific Name – *Myrmekioderma* sp.

Description – White, with an almost metallic appearance and convoluted surface. Encrusting. Genus name is based on resemblance to **Orange Convoluted Sponge** (*M. styx*) that has not yet been confirmed in Anguillian waters, so identification is only tentative. Uncommon, although is locally abundant around parts of Scrub Island.

Common Name – **Encrusting Star Sponges**

Scientific Name – Various

Description – Variable colours and low-lying with generally small ex-current openings. Some varieties have star patterns around these openings, while others do not, having instead slightly raised lips. Other species have both characteristics. Identification to species or even Genus is often unreliable and so are mentioned here generically. Opposite are a two examples of such species, with common names assigned based on their morphology: Red Star Sponge and White Star Sponge.

Common Name – **Orange Icing Sponge**

Scientific Name – *Mycale laevis*

Description – Bright orange with large, transparent, ex-current openings. Only a small portion of this species is usually visible as it encrusts under the margin of plate-like corals and ledges. Its presence usually modifies the growth pattern of the coral's edges. Occasional, although its cryptic nature means it is often overlooked.

Notes on other sponge species: Sponge morphology can vary greatly within a species making identification very precarious, and inter-species similarities can further confuse the issue. For example, small **Stovepipe Sponges** (*Aplysina archeri* - pictured right) can sometimes resemble the purple colour morph of **Branching Tube Sponge** (*Pseudoceratina crassa* - top of page 31). Interestingly, although this species is considered relatively common in some parts of the Caribbean, it has yet to be positively identified in Anguillian waters. There is such a variety of sponge species present that it is not possible to fully represent them all in this book. Two such unidentified examples are pictured below.

Two of the many as yet unidentified sponge species that can be found in Anguillian waters (photo credit Stuart Wynne)

3.2 Phylum – Cnidaria: Cnidarians

Cnidarians are relatively simple creatures, being generally small in size and grouping together to form colonies. The structure of these colonies varies greatly from the small unobtrusive hydroids (page 41-43) to arguably the most important creatures in a coral reef ecosystem, the hard corals (page 51-64). These latter colonies are sometimes massive and usually make up the bulk of the reef structure. Some cnidarians however are not colonial and live individually in open-water (Jellyfish – page 44) or attached to a hard surface by hydrostatic pressure (Anemones – page 45-48). Individual animals have a general cup-shaped body plan with an oral disc surrounded by tentacles and a central opening that serves as both mouth and anus. The majority of Cnidarians spend their lives as attached polyps (see photograph for example of classic polyp form) although many groups have a free-swimming stage that resembles a jellyfish, known as a medusa. Some groups however (e.g. the anthozoans – page 45-68) do not have this free-swimming stage, with polyps instead reproducing by releasing sperm and eggs simultaneously into the water. This process ultimately produces a short-lived free-swimming larvae (planula) that soon settles onto a suitable surface and develops into an adult. Further differences between groups are described later in this section.

Many cnidarian species share a highly significant symbiotic relationship with certain species of dinoflagelate algae known as zooxanthellae. These algae live in the surface tissue of the creature and photosynthesize, providing important nutrients to the 'host'. In return the algae profit from the host's metabolites that would otherwise be excreted as waste. In tropical seas where nutrients are in short supply this relationship is highly important. Indeed, during times of stress the hosts may inadvertently expel their guests and if this continues for a prolonged period they will die. The algae contain pigments that give their hosts specific colourations, so during this time the host becomes white, and thus the process is known as bleaching. This subject has become increasingly significant over recent years and is discussed in more detail on page 69-70.

The cnidarians get their name from the Latin word for 'nettle' because almost all members of this Phylum uniquely possess stinging 'capsules' called nematocysts. These small packages located mainly on their tentacles contain a coiled, pointed, harpoon-like structure, that when triggered is discharged as either a means to capture prey through paralysis or as defence. Most cnidarian stings have no harmful effect to humans, although a few can be quite toxic. If stung, never rub the infected area or wash with fresh water as this can discharge unspent nematocysts. Instead, pour vinegar over the area to immobilise remaining nematocysts and then sprinkle with meat tenderiser to alleviate pain. If in a remote area it is advisable to urinate on the affected area as this has a similar effect.

Interestingly, sea slugs prey on some species of hydroids and anemones, ingesting their nematocysts without discharging them and incorporate them into their skin, thus using the defence capabilities for themselves!

Class – Hydrozoa: Hydrozoans

The hydrozoans have the most varied body plans of all cnidarians. They are typically predatory animals, with some species forming colonies that are fixed to the sea bed and others that are free-floating/swimming colonies. One characteristic of a hydrozoan is that there is a division of labour between the various polyps in the colony. Some specialise in feeding, while others specialise in defence or reproduction. All polyps share nutrients through an interconnecting digestive tube.

Order – Milleporina: Fire Corals

Fire coral has a hard calcareous skeleton that appears relatively smooth, although on close inspection it is possible to see the tiny hair-like polyps that protrude from the surface. Fire coral are called such because they can produce a painful sting if touched due to unusually powerful stinging nematocysts. As with all corals, fire corals can be delicate and are easily broken so should not be touched even if gloves are worn for protection.

Common Name – **Branched Fire Coral**

Scientific Name – *Millepora alcicornis*

Description – The colony forms cylindrical structures with its skeleton usually golden in colour although the tips may be pale or white. It can also encrust other structures taking on their shape. Often seen growing in areas with lots of water movement. Common to abundant.

Common Name – **Blade Fire Coral**

Scientific Name – *Millepora complanata*

Description – The colony forms upright blades or plates that extend from an encrusting base, with its skeleton usually golden in colour although the edges may be pale or white. Often seen growing in high energy environments, sometimes on the tops of patch reefs in water so shallow that it becomes exposed in between tidal swells. Common to abundant.

Order – Stylasterina: Lace Corals

Lace corals have an intricately branched, hard calcareous skeleton. Like the closely related fire corals they have hair-like polyps that extend through pores in their skeleton, although the polyps form small cup-like structures that give the outer branches a serrated appearance. The stings of lace corals are not as severe as those from the fire corals although irritation to sensitive skin can still occur. However, like all corals, and especially the case here, lace coral is fragile and so contact should be avoided at all times.

Common Name – **Rose Lace Coral**

Scientific Name – *Stylaster roseus*

Description – Colonies are small and generally found in crevices and in other sheltered areas. They are usually lavender to pink in colour and considered locally uncommon, with the majority of reported sightings to date being made in areas near Sandy Island.

Order – Hydroida: Hydroids

Most species of hydroids are colonial and form feather or fan-like structures that are usually small and unobtrusive, so much so that they are often overlooked or confused with plants or algae. Because of their small size positive identification is difficult underwater, sometimes needing microscopic examination, and as such only the most distinctive varieties have been included here. Probably the most commonly noticed species is the Christmas Tree Hydroid (*Halocordyle disticha*) because it often grows on mooring ropes and has a mild sting that regularly makes one aware of its presence. Having said this, most species sting to some extent – especially on sensitive skin, with the most painful reportedly being the Feather Bush Hydroid (*Dentitheca dendritica*).

Common Name – **Branching Hydroid**

Scientific Name – *Sertularella speciosa*

Description – Branches extend alternately from a central stalk and have obvious white polyps attached on alternating sides. Can grow singly or in clusters and are found on a variety of surfaces from rocks to seagrass blades. The central stalk is relatively thick compared to the side branches. Considered common.

Common Name – **Christmas Tree Hydroid**

Scientific Name – *Halocordyle disticha*

Description – Branches extend alternately from a central stalk, each with an obvious white polyp at the end. Other polyps grow on the branches but are not as pronounced. There are sometimes algal growths among the branches. Colonies often grow in clusters, and are considered common although will often be overlooked due to their inconspicuous nature.

Common Name – **Thread Hydroid**

Scientific Name – *Halopteris carinata*

Description – Long, thin central stalk with short alternating side branches. Will grow on all reef areas, often in clusters. As with many hydroid species identification requires microscopic examination. Occasional.

Common Name – **Feather Bush Hydroid**

Scientific Name – *Dentitheca dendritica*

Description – Colonies form bush-like structures with thick central stalk and stout primary branches and sub-branches. These latter branches are lined with fine, tightly grouped secondary branches that house the inconspicuous polyps. These secondary branches look similar to other species that form less massive colonies, for example Feather Hydroids (*Gymnangium longicauda* – page 43). The Feather Bush Hydroids usually have Hydroid Zoanthids (*Parazoanthus tunicans* – page 49) encrusting their main branches.

Common Name – **Solitary Gorgonian Hydroid**

Scientific Name – *Ralpharia gorgoniae*

Description – Largest of the Caribbean hydroid polyps, these solitary individuals attach themselves to soft corals. The pictured specimen is attached to a sea fan. The soft coral polyps often grow up the hydroid's 'stem', which makes it difficult to discern where one species stops and the other species starts. Interestingly, this particular species was formerly considered rare in Anguillian waters until June 2009, when a number of specimens were sighted in different locations simultaneously. This raises the question of seasonality.

Notes on other hydroid species: It is quite likely that other species of hydroids live in Anguillian waters although their inconspicuous nature and similarities in appearance mean a full list has yet to be compiled. Candidates include the **Feather Hydroid** (*Gymnangium longicauda* - pictured right), **Algae Hydroid** (*Thyroscyphus ramosus*), **Unbranched Hydroid** (*Cnidoscyphus marginatus*) and **Stinging Hydroid** (*Macrorhynchia* sp.).

Order – Siphonophora: Siphonophores

Siphonophores (not pictured) can often be confused with jellyfish as they are transparent, free-swimming individuals. However, what looks like a single individual is in fact a colony of individuals that controls its buoyancy using a gas-filled float, hanging below which are numerous nematocyst-bearing tentacles. Siphonophores do not have a stage in their lifecycle that attaches to the sea bed or other structures, but are instead free-swimming. The colonies are generally small in size, only being noticed by swimmers at certain times of the year when they can congregate in shallower areas and cause mildly annoying stings. These stings are usually short-lived, however one member of this group is much larger and more fearsome: the **Portuguese Man-O-War** (*Physalia physalis*). Although not reported in Anguillian waters this siphonophore is pink to purple in colour and floats on the surface with a large gas-filled float that also acts like a sail, propelling the submerged colony through the water. The tentacles can drift behind it for quite a distance and remain toxic for some time if washed up on a beach. Any sightings of this species should be reported to the Department of Fisheries and Marine Resources as soon as possible and no attempt be made to capture it.

Order – Leptothecatae: Hydromedusa

The Hydromedusa are often confused with juvenile jellyfish as they have an almost identical body structure, if somewhat miniaturised. They are a type of non-colonial free-living hydroid whose dominant life phase is that of the medusa, in some ways similar to the siphonophores. Many species are slightly toxic, but only affect those with sensitive skin. Beginning around the month of May many thousands of these creatures can be seen in coastal areas, persisting until September/October.

Class – Scyphozoa: True Jellyfishes

The true jellyfish are translucent unattached medusae that swim in open-water, and in many ways are similar to the hydromedusa (page 43). All have a dome-shaped 'body' that varies from a shallow bowl shape to a deep bell, which is surrounded on its margin by many nematocyst-bearing tentacles. The mouth is in the central underside of this dome, and in some species four frilly oral arms hang from this feeding tube that also carries nematocysts. Jellyfish use pulsating contractions of their dome to propel them through the water, and although only a few species are toxic, caution should generally be taken with this Class.

Reproduction of a typical jellyfish is an interesting variation on the cnidarian theme. Males and females release gametes into the water that subsequently fuse and develop into larvae (known as a planula). The planulae eventually attach to a rock or other hard surface, at which point they become known as a polyp, later 'budding' a replica polyp underneath them. This new polyp follows suit and the process continues, slowly forming a 'colony' that resembles a stack of saucers. The polyps in the stack gradually develop tentacles, and eventually the uppermost individuals break off and become new free-swimming medusa. These medusa finally grow into a sexually mature adults.

Common Name – **Moon Jelly**

Scientific Name – *Aurelia aurita*

Description – Saucer-shaped dome with numerous short, fine tentacles around its margin. Dome is translucent but their reproductive organs, shaped like a four-leafed clover, can be clearly seen through it. Can have a pink tinge, and may form huge groups of individuals in shallow to deep water from early May until September/October.

Common Name – **Upsidedown Jelly**

Scientific Name – *Cassiopea* sp.

Description – Flattened dome that becomes ring-like while pulsating. Of the two species present in the region exact distributions and abundances of each are unknown, with both being physically very similar. They inhabit calm areas, for example back reefs, lagoons and mangrove areas - the paucity of these kinds of habitat in Anguilla probably explains the small number of sightings reported. Both species have the curious habit of swimming upside down, pressing their dome flat to the sea bed and filter feeding with their tentacles. Such a habit may well explain how anemones came into being (page 45-48). They also share a symbiotic relationship with algae known as zooxanthellae that grow in their tissues. In a similar way to some other cnidarians they provide the jelly fish with part of their nutrient requirements.

Class – Anthozoa: Anthozoans

Anthozoans differ from many other cnidarian groups because they don't have a free-swimming stage in their reproductive cycle. Instead, the polyps release sperm and eggs consecutively that fuse and become larva (known as a planula), and in much the same way as a jellyfish planula, it settles on the bottom and becomes a polyp. This polyp buds new polyps and forms a colony, but unlike the jellyfish colony, medusa do not form and swim away, instead the colony grows, and continues to do so sometimes reaching mammoth proportions as with the hard corals (page 51-64). There is much variety within the anthozoans, some being quite large solitary polyps, while others are small colonial types. Some are completely soft-bodied, whilst others build hard calcareous skeletons that are ultimately responsible for creating the vast majority of the coral reefs structure. However, all species have tentacles that they use predominantly to filter feed plankton or small fish and invertebrates from the water.

The anthozoans described in this book are divided into two main groups: Hexacorals and octocorals. The hexacorals have tentacles arranged in multiples of six, although this arrangement is not always obvious due to the sheer number of tentacles. Group members include the anemones, zoanthid, corallimorphs, hard corals and black corals. On the other hand, octocorals have eight tentacles, and this tentacle arrangement is easier to spot as members bear far fewer tentacles, often only eight. They are represented in this book by the soft corals (page 65-68). This arrangement can in fact be a useful identification feature when trying to distinguish a few species of larger hydroids and smaller zoanthid colonies from the octocorals.

As an interesting note, the black corals are considered in a different taxonomic group from other hexacorals because of their very different body plan (page 64). However, because each polyp possesses six tentacles they are still considered hexacorals.

Order – Actiniaria: Sea Anemones

Sea anemones are soft-bodied animals that lack any hard skeletal parts. No species are colonial although some may live close to one another, and all live attached to a variety of surfaces where they filter feed with their tentacles. These tentacles possess nematocysts, although very few are powerful enough to be felt by humans. They are however toxic enough to paralyse their prey, usually consisting of small fish or invertebrates, which the tentacles then manoeuvre towards the mouth situated in the centre of the oral disc. The tentacles vary in size, shape and number, often being a key identification feature. For example, their arrangement usually appears random, distinguishing them from corallimorphs (page 50) which have a much more orderly pattern.

Some creatures are immune to their stings and live in close association with anemones, using their toxicity as a defence of their own. Although popularised in the cinema in recent years, the most famous of these associations does not occur in the Caribbean, as clown fish are only found in Indo-Pacific coral reef regions. Instead, similar associations occur with other species of fish, and some types of shrimp and crab.

Common Name – **Giant Anemone**

Scientific Name – *Condylactis gigantea*

Description – The largest anemone in Anguillian waters with long, chunky tentacles that have a slightly swollen tip. The main body is usually white, with the tentacles becoming pink towards the tip. Colourations may vary, with green to tan tentacles as in the pictured examples (right and above inset). Considered occasional, although is often overlooked as can live in reef recesses or under rocks. May produce slight skin irritation on very sensitive skin types.

Common Name – **Sun Anemone**

Scientific Name – *Stichodactyla helianthus*

Description – Short pale green to brown tentacles with no obvious central oral disc. Usually found in shallow reef areas, although considered occasional in Anguilla. Has been seen in Sile Bay, Maundays Bay and along Seal Island Reef System.

Common Name – **Elegant Anemone**

Scientific Name – *Actinoporus elegans*

Description – Short knobby tentacles cover the flat oral disc that has white and black ladder-like patterning. Usually found on areas of sand or seagrass. Pictured specimen was seen close to Scilly Cay. Rare.

Common Name – **Corkscrew Anemone**

Scientific Name – *Bartholomea annulata*

Description – Has many long thin translucent tentacles that are marked with a white spiral pattern that resembles a corkscrew. Body is greyish in colour although is often hidden under a rock. Considered occasional, although is often overlooked because of its cryptic habits. May produce slight skin irritation on very sensitive skin types.

Common Name – **Beaded Anemone**

Scientific Name – *Epicystis crucifer*

Description – Frilly oral disc with densely packed relatively short tentacles around its edge in fairly distinctive rows. The tentacles look quite swollen, sometimes with bar-like raised areas across their width. Bead-like warts radiate from the mouth to the tentacles. Uncommon.

Common Name – **Hidden Anemone**

Scientific Name – *Lebrunia coralligens*

Description – Lives in cracks with only the enlarged ends of long false-tentacles protruding. These false-tentacles are usually creamy white in colour with brown tips. The true tentacles are usually only extended at night. Considered uncommon, although its cryptic nature means a true abundance is difficult to assess.

Common Name – **Branching Anemone**

Scientific Name – *Lebrunia danae*

Description – Has a network of psuedotentacles which hide the body from view and are made up of thin, branched, pointed sections and swollen, round nematocyst-bearing knobs. These structures are usually shades of brownish red on one side and pale to white on the other. Occasional, with most current sightings being made around Sandy Island. Their true tentacles are only extended at night and its nematocysts are mildly toxic, stinging bare skin.

Common Name – **Sponge Anemone**

Scientific Name – Unidentified

Description – Translucent brown tentacles are marked with white bands, with the oral disc often having a white central portion as in the pictured specimen. Lives in association with sponges. Can be solitary but often lives in pairs or small groups. Is relatively flat and although considered common, its exact scientific identification has yet to be established.

Notes on other anemone species: A great many anemones in Anguillian waters remain unidentified, perhaps due to the great diversity present, but also due to the fact that there are very few taxonomists presently working on this fascinating group of animals.

Order – Ceriantharia: Tube-Dwelling Anemones

Tube-dwelling anemones are closely related to regular anemones, but differ due to a parchment-like tube that they retract into when disturbed. Many species are nocturnal, only emerging from their tube to feed at night, and most live in areas of sand or rubble. Unfortunately, there are currently no known taxonomists working on this Order in the tropical western Atlantic and so many species lack a scientific description. The specimen pictured right is probably a **Banded Tube-Dwelling Anemone** (*Arachnanthus nocturnus*), with translucent brown-white banded outer tentacles, a white oral disc and whitish central tentacles. There are however several similar species and positive identification requires microscopic examination by an expert.

Order – Zoanthidea: Zoanthids

Zoanthids look like miniature anemones, also bearing a resemblance to coral polyps. They differ from anemones because they usually live in colonies or in close association with others of the same species, and have an oral disc that lacks tentacles apart from two rings around its margin. This tentacle arrangement is similar to that of many coral species although zoanthids lack a rigid colonial skeleton. Some of the smaller species live in close association with sponges, hydroids and other invertebrates.

Common Name – **Sponge Zoanthid**

Scientific Name – *Parazoanthus* sp.

Description – Small polyps that form large groups of individuals on a variety of sponge species (for example: Branching Vase Sponge *Callyspongia vaginalis* and Lavender Rope Sponge *Niphates digitalis* – page 31 & 34 respectively). Colours vary, but are often yellow to orange. Considered abundant, providing there are host sponge species present.

Common Name – **Golden Zoanthid**

Scientific Name – *Parazoanthus swiftii*

Description – Relatively large golden-yellow individuals encrust the surface of several sponge species that are often encrusting in nature themselves (as in the pictured specimen). Considered uncommon.

Common Name – **Hydroid Zoanthid**

Scientific Name – *Parazoanthus tunicans*

Description – Relatively large individuals which encrust Feather Bush Hydroids (*Dentitheca dendritica* – page 42), usually pale yellow-brown in colour. Considered uncommon and limited in range by abundance of hydroid 'host', which in the pictured example was found growing from a rock on the sand flats beyond Limestone Bay reef.

Common Name – **Mat Zoanthid**

Scientific Name – *Zoanthus pulchellus*

Description – Has relatively thick tentacles, and grows in mat-like groups that are sometimes so dense the round polyps take on a more angular shape. Usually green in colour, although this can vary depending on depth. Considered uncommon with most recent confirmed sightings being in parts of Shoal Bay East.

Common Name – **White Encrusting Zoanthid**

Scientific Name – *Palythoa caribaeorum*

Description – Looks relatively hard, and the closed polyps may be mistaken for Great Star Coral (*Montastraea cavernosa* – page 55) by the untrained observer. In fact it is spongy in texture, and when the polyps are extended this confusion is easily avoided. Colouration is usually golden brown, although may become paler especially if undergoing bleaching (page 69). Is considered common although can be locally abundant especially in shallower reef areas.

Order – Corallimorpharia: Corallimorphs

Corallimorphs are very similar in appearance to anemones (page 45-48), although there are a couple of general differences which are key to correctly distinguishing them. Firstly, the oral disc is usually quite flat, with a mouth that protrudes from the surface and surrounded with short, stubby tentacles (although in some species they are almost absent). Secondly, the tentacles of corallimorphs, unlike most anemones, do often follow a pattern and form concentric rings around the mouth, with tentacles forming lines that radiate out somewhat like spokes. This pattern can be obscure so a third key feature, the fact that corallimorphs, although sometimes solitary, often live in close association with others of the same species, should also be used to differentiate the two groups. Despite this, even though anemones are considered solitary, they can sometimes be found in groups. This means that identification between groups, although possible, needs careful consideration.

Common Name – **Florida Corallimorph**

Scientific Name – *Ricordea florida*

Description – Short, swollen, knob-like tentacles cover the oral disc, becoming longer towards the margin. When the disc is fully open it lies flat and reveals the tell-tale radial spoke-like tentacle arrangement. Usually green in colour but does vary, often with almost florescent yellows and blues. Non-toxic. Uncommon.

Order – Scleractinia: Hard Corals

The hard corals, sometimes referred to as the stony corals, are the basic building blocks of tropical coral reefs. They are sensitive creatures that live in a relatively narrow temperature, pH, nutrient, and light intensity range. Variables other than these can also affect coral mortality, and this subject has become of increasing importance over recent years and is discussed in detail on page 69.

Hard coral polyps grow in sometimes vast colonies and secrete a calcium carbonate skeleton that takes on the generalised shape of a small cup (known as the corallite), inside which the polyp sits protected. The tentacles of the polyp are usually retracted during the day, only being visible at night, although in some species they are extended at all times unless molested. Others are solitary, or only grow in association with other polyps, thus not forming traditional colonies. Colonies increase in size by budding replicates next to themselves with successive generations overgrowing each other. Colony size and shape varies between species, with some that can be recognised purely by the growth forms of the colony or solitary polyp. Other species, however, look very similar to one another from a distance, and so the pattern of corallite 'spokes', known as septa, may need to be examined to ensure correct species identification. Having said this, identification can be quite tricky for certain species when close examination of shape and form of corallites is necessary, especially when underwater!

The hard corals listed in this book have been arranged by shape and general appearance, rather than solely using traditional scientific classification. The main grouping criterion is corallite structure and how it relates to colony growth forms. For example, there are two main corallite growth patterns: one being the typically circular cup-like structure mentioned above, the other being a valley-like grooved structure where corallites have fused together. Such growth form variation means that a hard coral will ultimately fall into one of four general categories: Colonial cup-like corallites; colonial brain-like corallites; colonial sheet-like corallites; or solitary cup-like corallites. There is another non-taxonomic division that can be made, which is based on whether a coral species is considered a 'reef builder', i.e. one that deposits substantial amounts of calcium carbonate to the reef structure. These are known as hermatypic corals, and make up the majority of this section. The ahermatypic corals are those that do not contribute substantially to the deposition of calcium carbonate onto the reef structure, and are mainly represented here by the solitary cup corals.

Common Name – **Staghorn Coral**

Scientific Name – *Acropora cervicornis*

Description – Colonies form antler-like arrangements that can grow into tangled structures. Colony surface is rough with protruding circular corallites. Prefers shallow depths in clear, calm water on reefs although will sometimes grow on sand and is usually a pale tan to gold colour. Once abundant in the Caribbean this species was devastated by white-band disease (page 69) in the late 1980s and early 1990s is now considered uncommon to rare. It is a rapid grower but fragile and so, as with all corals, contact should be avoided.

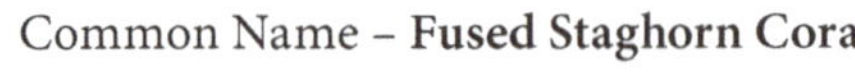

Common Name – **Fused Staghorn Coral**

Scientific Name – *Acropora prolifera*

Description – Molecular evidence suggests that this species is in fact a hybrid between Staghorn Coral (previous) and Elkhorn Coral (below). Forms short antler-like growths that often fuse into flattened paddle-like structures. Uncommon. It is a rapid grower but fragile and so, as with all corals, contact should be avoided.

Common Name – **Elkhorn Coral**

Scientific Name – *Acropora palmata*

Description – Colonies form flattened branches that can grow into large impressively beautiful structures. Their surface is rough with protruding circular corallites that are often pale to white in colour whereas the rest of the colony is pale tan to gold. Once abundant in the Caribbean this species was devastated by white-band disease in the late 1980s - early 1990s (page 69-70), and is now considered occasional in Anguillian waters although it is still common in some locations. It is a rapid grower but fragile and so, as with all corals, contact should be avoided.

Common Name – **Finger Coral**

Scientific Name – *Porites porites*

Description – Colonies form small finger-like branches, sometimes covering large areas, with various growth forms that are considered by some as different species. However, for the purpose of this book they will be considered a single species. Silver grey in colour the colonies can have short stubby branches, longer quite chunky branches, or long slender branches. There is also a growth form of the long slender variety that is lavender in colour. Common on most reef habitats although remains of dead colonies suggest abundance was historically higher. Some colonies alive today appear to be slowly dying back, either breaking or being smothered by algae. It is a rapid grower but fragile and so, as with all corals, contact should be avoided. Appears somewhat similar to Ten-Ray Star Coral (page 54).

Common Name – **Pillar Coral**

Scientific Name – *Dendrogyra cylindrus*

Description – Colonies form thick spires that grow upwards and look somewhat similar to stalagmites. Can be massive and elegant-looking. Colony surface has a fuzzy appearance, almost like a Persian rug, caused by the polyp's tentacles that are almost always extended. Fallen pillars can give rise to several new upright spires. Colour usually light tan to gold. Occasional.

Common Name – **Yellow Pencil Coral**

Scientific Name – *Madracis mirabilis*

Description – Colonies form densely packed groups of pencil-like structures that can look similar to Eight-Ray Finger Coral (next). Usually yellow to green in colour, and sometimes forming huge colonies. The polyp tentacles are often protruding during the day giving a soft appearance. It is a rapid grower but fragile and so, as with all corals, contact should be avoided. Uncommon, although considered occasional in some locations, for example Limestone Bay and parts of Shoal Bay East.

Common Name – **Eight-Ray Finger Coral**

Scientific Name – *Madracis formosa*

Description – Colonies can form dense finger-like groups that appear quite similar to Yellow Pencil Coral (previous) although they are less fragile-looking with thick, often double-lobed tips. Yellow to green in colour, with most corallites having eight septa. Although its name suggests otherwise, some corallites have ten septa, thus overall growth form is a better identification method. The polyp tentacles are often protruding during the day giving a soft appearance. Uncommon.

Common Name – **Ten-Ray Star Coral**

Scientific Name – *Madracis decactis*

Description – Colonies appear quite similar to Finger Coral (*Porites porites* - page 52), although are green to yellowish-tan-grey in colour and generally more encrusting in nature, with raised, often densely packed lobes. Corallites have ten septa and the polyp tentacles are often protruding during the day giving a soft appearance, although in the pictured specimen they are retracted. Uncommon.

Common Name – **Blushing Star Coral**

Scientific Name – *Stephanocoenia intersepta*

Description – Colonies form generally smooth low-lying domes not usually more than 20 cm in diameter. The skeleton of the colony is light orange in colour but the polyps inside the corallite are dark. When partially extended during the day a waft of water will cause the polyps to retract, paling the colony significantly. As the polyps extend again the colony appears to blush, hence its common name. The septa are clearly visible and colonies will often have tubeworms burrowed into them, their filter feeding apparatus extended. Occasional on most reef habitats.

Common Name – **Orbicella complex**

Scientific Name – *Orbicella* sp.

Description – Colonies grow in different configurations, with three similarly appearing species. For the purpose of this book these three species are described under the generic heading of Orbicella complex. Firstly, most common in Anguillian waters is **Lobed Star Coral** (*Orbicella annularis*) pictured left. This forms green to yellowish-brown lobe-shaped colonies on the tips of protruding bumps of sometimes enormous coral heads - presumably originating from one colony but not now connected. Corallites are fairly smooth and uniform in size, and the colony margin does not overhang.

Boulder Star Coral (*Orbicella franksi*)

Mountainous Star Coral (*Orbicella faveolata*)

Second most common is **Boulder Star Coral** (*Orbicella franksi)* pictured above left, which is similar in colour but colony margins overhang underlying rocks. The colony surface is covered with randomly distributed raised lumps and corallites that are unevenly distributed and vary in size. Finally, **Mountainous Star Coral** (*Orbicella faveolata*) pictured above right, is similar to Boulder Star Coral although the raised lumps are arranged in lines, thus resembling a mountain range, and the corallites are more equally spaced and uniform in size. Note: These three species were previously classified in the Genus *Montastraea*.

Common Name – **Great Star Coral**

Scientific Name – *Montastraea cavernosa*

Description – Colonies form small to large boulders and domes although sometimes may take on a more encrusting form. Polyps and their corallites are large, bulbous structures and very distinctive of this species. Colours vary from green to brown to yellow-grey. Common on most reef environments. Not to be confused with White Encrusting Zoanthids (*Palythoa caribaeorum*) on page 50.

Great Star polyp detail

55

Common Name – **Smooth Star Coral**

Scientific Name – *Solenastrea bournoni*

Description – Colonies form golden domes that usually have irregular bumps on surface. Centres of corallites are dark with their edges protruding slightly, giving them a bumpy appearance. Prefers shallower reefs where it is considered uncommon to rare.

Common Name – **Elliptical Star Coral**

Scientific Name – *Dichocoenia stokesi*

Description – Colonies are yellow to brown in colour and form rounded domes, which are usually small in size. Corallites are relatively large, protrude noticeably, and squash together with neighbouring polyps to form elliptical or sometimes Y-shaped colonies. This latter feature distinguishes it from the similar appearing Golf Ball Coral (next), also usually forming larger colonies. Occasional on most reef habitats.

Common Name – **Golfball Coral**

Scientific Name – *Favia fragum*

Description – Colonies are yellow to brown in colour and form rounded domes, which are usually small in size. Corallites are relatively large, but do not protrude noticeably and are generally well spread out. This latter feature, together with the fact that colonies are generally smaller, distinguishes it from Elliptical Star Coral (previous). Uncommon on most reef habitats.

Common Name – **Mustard Hill Coral**

Scientific Name – *Porites astreoides*

Description – Colonies are usually domed and lumpy, resembling a cauliflower. Can sometimes encrust, although its lumpy appearance remains. Colour varies from a vivid green to yellow to tan-grey. Usually relatively small and probably the most common species of coral in Anguillian waters, especially in shallower areas. Abundant on most reef habitats.

Common Name – **Massive Starlet Coral**

Scientific Name – *Siderastrea siderea*

Description – Colonies usually domed, sometimes large and even spherical. Has a smooth surface with indented corallites that resemble the marks left when a pencil is poked into putty as the septa are not clearly visible (below) – this differentiates it from the similar Lesser Starlet Coral (next) that has more pronounced septa. Colony colour varies from green to yellowish-brown or tan. Common on most reef habitats.

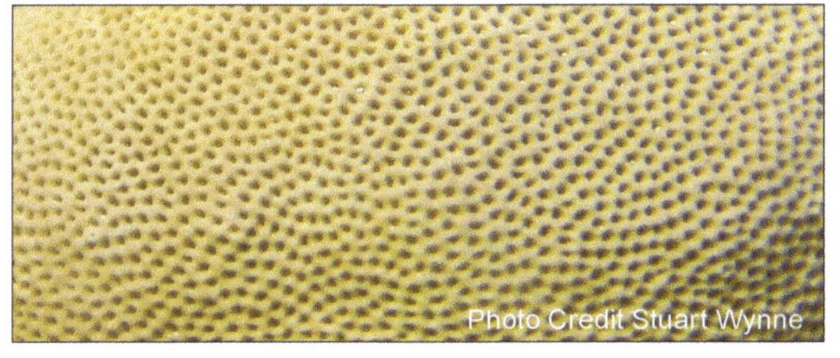

Common Name – **Lesser Starlet Coral**

Scientific Name – *Siderastrea radians*

Description – Colonies are small and resemble Massive Starlet Coral (previous) although septa are far more pronounced. Lesser Starlet Coral is one of the few species that can thrive in sediment-rich water, and are thus sometimes the only species present in these areas, even colonising rocks in shallow turbid coastal conditions. Usually pale tan in colour, but may sometimes be almost white. Abundant on most reef habitats.

Common Name – **Symmetrical Brain Coral**

Scientific Name – *Diploria strigosa*

Description – Colonies have long ridge and valley patterns that resemble a brain. Colour varies from green to yellowish-brown and colonies are usually domed, although may sometimes encrust. This species can be differentiated from the similar Boulder Brain Coral (next) because its ridge and valley pattern does not usually form enclosed structures and is generally less massive. The ridges also lack an obvious line along their top, although a faint one is sometimes present. Common on most reef habitats.

Common Name – **Boulder Brain Coral**

Scientific Name – *Colpophyllia natans*

Description – Colonies have long ridge and valley patterns that resemble a brain. Colour varies from green to yellowish-brown with the valleys usually lighter or of contrasting colour. Although this species looks similar to the other brain corals, it belongs to a different Genus, with the main visual difference being that Boulder Brain Coral has an enclosed valley and ridge structure, with a clearly indented line along the ridge tops (see top left image on following page). It is also generally larger, forming wider ridges and valleys and often massive domes, although smaller colonies are more common in Anguilla. Occasional on most reef habitats.

Intricate patterns of Boulder Brain Coral. Note indented ridge line in the left hand image.

Common Name – **Knobby Brain Coral**

Scientific Name – *Diploria clivosa*

Description – Colonies are somewhat dome-shaped, although sometimes encrust, and have long interconnected ridges and valleys that resemble the classic brain pattern. It can be distinguished from the other brain-like corals by numerous irregular knobs that bulge from its surface, although these can be indistinct, meaning it resembles Symmetrical Brain Coral (two species previous). A key feature is that the ridges of Knobby Brain Coral are usually lower, more rounded, and with a somewhat fuzzy appearance caused by partially extended polyp tentacles. Colony colour varies from green to yellowish-brown although the valleys are usually lighter or of contrasting colour. Occasional on most reef habitats.

Common Name – **Grooved Brain Coral**

Scientific Name – *Diploria labyrinthiformis*

Description – Colonies usually form domes and have long interconnected ridges and valleys that resemble the classic brain pattern, although the valleys of this species take two forms: one is the polyp-bearing 'true valley' of the colony, whereas the other is a 'false valley' that creates a noticeable divide. This latter groove can vary in width, and sometimes may only just be visible making identification problematic. Other colonies have grooves that may be considerably wider than the true valleys. Colony colour varies from yellowish-tan to brownish-grey. Common on most reef habitats.

Common Name – **Maze Coral**

Scientific Name – *Meandrina meandrites*

Description – Colonies form domes although sometimes appear as flattened plates. Looks somewhat similar to the brain corals (previous) although resembles sharp, jagged, piped cake icing rather than a brain. The valleys house the polyps that live in long lines of connected corallites with pointed ridges. Colonies are tan to yellow to almost bluish white in colour. Occasional on most reef habitats.

Common Name – **Rose Coral**

Scientific Name – *Manicina areolata*

Description – Colonies usually form small elliptical domes with a central valley and several short side ridges. This arrangement resembles the classic profile schematic of animal cell mitochondria. Colour varies greatly, although usually uniformly pale. Uncommon on reef habitats but is one of the only species of coral commonly found growing within seagrass beds.

Common Name – **Whitestar Sheet Coral**

Scientific Name – *Agaricia lamarcki*

Description – Colonies encrust rocks or form thin sheets with concentric rows of long ridges. The valleys contain white star-like corallites, with the surrounding structures golden-brown in colour. Inhabits a variety of reef areas although is considered uncommon in Anguilla.

Common Name – **Lettuce Coral**

Scientific Name – *Agaricia agaricites*

Description – Colonies usually encrust rock sometimes forming upright plates or lobes. Has ridges of different heights and thicknesses that resemble the surface of an iceberg lettuce or cabbage. May often be found growing in reef recesses or crevices. Colony colour varies from tan to yellowish grey-brown. Small colonies closely resemble Low Relief Lettuce Coral (next). Occasional on most reef habitats.

Common Name – **Low Relief Lettuce Coral**

Scientific Name – *Agaricia humilis*

Description – Colonies form small, often circular, bumpy encrustations of densely packed corallites. Usually tan to yellow in colour and almost indistinguishable from small colonies of regular Lettuce Coral (previous), so much so that some scientists consider it the same species, classifying it as '*form humilis*'. Corallite depressions are quite deep and surrounded by a sharp ridge that never forms long valleys. Considered occasional on most reef habitats.

Common Name – **Cactus Coral**

Scientific Name – *Mycetophyllia* sp.

Description – Colonies form flat, often round, plates with a soft external appearance. Their internal skeleton is very spiny, which is probably where the common name derives. Colonies are made up of a series of rounded ridges and valleys somewhat similar to the brain corals, but valleys are noticeably wider and ridges appear fleshy. Colony colour varies greatly and species within this Genus are often very difficult to differentiate. Pictured specimen is probably **Rough Cactus Coral** (*Mycetophyllia ferox*). Occasional on most reef habitats, but more common in deeper areas.

Common Name – **Sinuous Cactus Coral**

Scientific Name – *Isophyllia sinuosa*

Description – Colonies form small, almost spherical, domes with fleshy ridges and deep narrow valleys. Colour varies but is usually light brown to yellow with occasional iridescence. Prefers shallow reef environments and is often associated with areas of high sedimentation where it is considered occasional.

Common Name – **Solitary Disc Coral**

Scientific Name – *Scolymia* sp.

Description – These corals do not form colonies, instead the polyps form single, often large (up to 15 cm) fleshy and round polyps. The septa are only visible when the polyp has died and the skeleton is exposed. Central area of polyp is usually flat with small warty growths. The three most common species in the Caribbean are very similar and sometimes require microscopic examination of the corallite structure to confirm identification, so for the purpose of this book, they have been grouped together. Colours vary greatly, although in Anguilla most specimens found are green to yellowish tan, suggesting they are either **Artichoke Coral** (*Scolymia cubensis*) or **Solitary Disc Coral** (*Scolymia wellsi*). Occasional on most reef habitats although may often be overlooked due to their small size and cryptic nature.

Common Name – **Spiny Flower Coral**

Scientific Name – *Mussa angulosa*

Description – Colonies form clusters of polyps that look very similar to the Solitary Disc Corals (previous). The polyps are large and fleshy, with a somewhat rough texture. The corallite is made up of numerous spiked septa, hence its common name, although these are only visible after the polyp dies. Colony colour varies but is usually greenish-grey to blue. Rare on most reef habitats.

Common Name – **Smooth Flower Coral**

Scientific Name – *Eusmilia fastigiana*

Description – Despite its common name, colony corallites are sharp and jagged and look like a blob of piped icing. Corallites are round and grow on a stalk that separates them from others in the colony although on occasion two may be fused together. Colour varies greatly from yellow-brown to blue-green. Common on most reef habitats although small colonies will often be overlooked because of their cryptic nature

Common Name – **Orange Cup Coral**

Scientific Name – *Tubastraea coccinea*

Description – Colonies form clumps of quite small, brilliant orange, obviously cup-like corallites, which appear domed and fleshy when the polyp's tentacles are not extended. Like other cup corals (see following) this species does not posses zooxanthellae and is not considered a reef builder (ahermatypic). It is in fact an introduced species from the eastern Pacific or tropical Indo-Pacific region and the only such species of invasive coral known in Anguillian waters. It probably originally arrived in the Caribbean transported on a ship's hull, as the first specimens reported were found on vessels in the Netherlands Antilles in 1943. They are not considered invasive however as they do no known harm to native species. Seen in abundance on the M.V. Catheley H wreck and also under various overhangs, although is considered occasional in Anguilla.

Notes on other cup coral species: Cup corals in the Caribbean predominately belong to two Families: Dendrophylliina, of which Orange Cup Coral (*Tubastraea coccinea* – page 63) is a member; and Caryophylliidae, of which Smooth Flower Coral (*Eusmilia fastigiata* – page 63) is a member. Other than these two species most other cup corals appear solitary, although they are in fact connected to neighbouring polyps by a rudimentary skeleton that is usually covered by encrusting sponges or other organisms. Corallites are cup-like with pronounced septa, but intra-species colour diversity means identification is often problematic. This, coupled with the fact that most species are difficult to observe because they live under overhangs or in caves, means that species diversity and abundance in Anguillian waters has yet to be established. Although these Cup Corals belong to the hard coral group, they are ahermatypic (non-reef building), as a true colonial structure is lacking. The pictured species is an unconfirmed cup coral, but probably either **Speckled Cup Coral** (*Rhizosmilia maculata*) or **Lesser Speckled Cup Coral** (*Colangia immersa*).

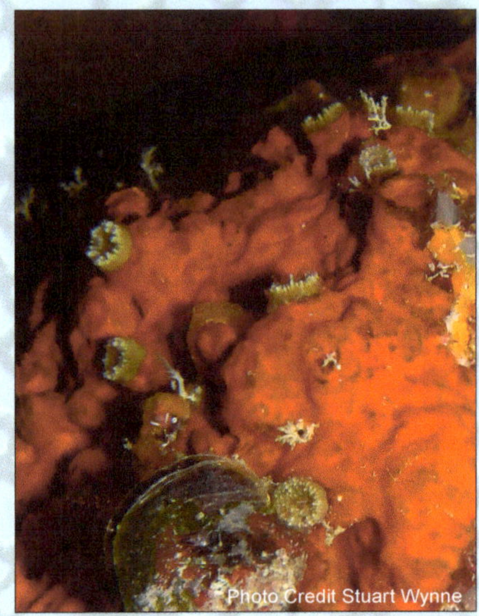

Unconfirmed cup coral species

Order – Antipatharia: Black Corals

Black corals are generally thought of as deep-dwelling species, although approximately half have been recorded within safe-diving limits. In many parts of the world larger species are often over-harvested, their polished skeleton being highly prized in the jewellery trade. This is not known to take place in Anguilla. The lack of black corals is probably due to a paucity of suitable deep, high relief, rocky habitat. Some specimens have reportedly been observed at Authors Deep dive site (see figure 1.2), although no photographic proof has yet been obtained.

Wire Coral (*Cirrhipates leutkeni*) seen at Dog Island

The polyps of black coral lay down a protein material in concentric rings that is black in colour and extremely hard, ultimately forming colonial shapes that are branch-like and thus similar to some of the octocorals. A key to differentiating black corals from octocorals is that the polyps of black coral live on the surface of this skeleton, each with six, non-retractable tentacles. Although the tentacles can expand and contract to some degree, they are usually visible to the naked eye.

Of the larger species confirmed in Anguilla **Feather Black Coral** (*Antipathes pennacea*) and **Wire Coral** (*Cirrhipathes leutkeni* - pictured right) are the species most likely to be seen by divers. Black corals favour deeper regions so it is rarely seen while snorkelling.

Order – Gorgonacea: Soft Corals

The gorgonians, usually typified by the classical sea fan, is in fact quite a varied group of what are commonly referred to as the soft corals. This term is in fact technically incorrect as soft corals are a large group of soft-bodied corals lacking any supporting skeleton. As this group is only represented by a couple of deep-water species in the Caribbean, their common name is often used to refer to the gorgonians. Gorgonians do however have a firm, yet flexible skeleton and so are not truly 'soft', which allows colonies of polyps to sway about in the water. This skeleton is made of a horny substance known as gorgonin which is laid down in concentric rings as the colony grows. All such 'soft corals' belong to the octocoral Subclass as each polyp bears eight tentacles. This is in contrast with the other coral groups which belong to the hexacoral Subclass as they have tentacles arranged in groups of six. The octocoral polyps live inside the usually tubular skeleton and extend their tentacles and upper bodies out of skeletal apertures to filter feed. The rim of the aperture is known as the calyx (plural calyces) and is often the key feature to species identification when combined with growth types. For this reason the following section is mainly split into colony growth types rather than species which, although not a traditional scientific classification, in most cases keeps members of the same Family and Genus together.

Common Name – **Encrusting Gorgonian**

Scientific Name – *Erythropodium caribaeorum*

Description – Flat, encrusting colonies with extended polyp tentacles that look like fine hairs. When polyps are retracted the skeletal surface, known as the rind, looks smooth and leathery. The polyps are pale tan colour and the rind a pale reddish tan, often with white aperture holes that contract and almost disappear when the polyp is retracted. The rind colour is a reliable feature to distinguish this species from the encrusting form of Corky Sea Finger (next). Common.

Common Name – **Corky Sea Finger**

Scientific Name – *Briareum asbestinum*

Description – Colonies usually form one or more erect structures somewhat reminiscent of fingers, although encrusting growth forms are sometimes observed. Tentacles are pale tan to light green in colour and when extended look like fine hairs. The calyces are somewhat swollen and the skeletal surface, known as the rind, is a purple colour – a reliable feature to distinguish the encrusting form of this species from the Encrusting Gorgonian (previous). Common.

Common Name – **Sea Rods**

Scientific name – *Pseudoplexaura, Plexaura, Eunicea, Plexaurella & Muricea* sp.

Description – Cylindrical, usually chunky skeleton, that in most species has only a modest number of branches. The Genus *Pseudoplexaura* consists of the Porous Sea Rods, whose polyp apertures have smooth calyces and are somewhat obscure. Some species of *Plexaura* look rather similar to the sea fans, while others do not (e.g. **Black Sea Rod** – *P. homomalla* - below left). *Eunicea* on the other hand is sometimes referred to as **Knobby or Warty Sea Rods**, as their calyces are somewhat swollen, especially the **Swollen-Knob Candelabrum** (*E. mammosa*) - below right. The *Plexaurella* are known as the Slit-Pore Sea Rods and, as their name suggests, they have pores that are slits rather than the typically rounder forms (e.g. **Giant Slit-Pore Sea Rod** – *P. nutans* - pictured right). Finally the Genus *Muricea* consists of members that have cylindrical calyces protruding significantly from the skeletal surface with jagged ends, often referred to as the spiny sea rods. Distinguishing species within these Genera is sometimes very difficult and requires microscopic identification. Sea rods are usually abundant on most reef habitats although density varies greatly from area to area, largely being influenced by intensity of wave action.

Giant Slit-Pore Sea Rod (*Plexaurella nutans*).

Sea Rod Polyp Detail.

Black Sea Rod (*Plexaura homomalla*).

Swollen-Knob Candelabrum (*Eunicea mammosa*).

Common Name – **Sea Plumes**

Scientific Name – *Muriceopsis* & *Pseudopterogorgia* sp.

Description – Thin, almost bushy, colonies with many branches on a single plane that originate from a collection of central 'stems'. The sea plumes are probably the most graceful underwater coral colonies, second to only large sea fans, as they sway around in ocean currents like Weeping Willow trees. Of the two Genera listed here the *Pseudopterogorgia* are the most elegant and common members of the group. They are usually branched on a single plane and can grow to quite impressive sizes, sometimes over two metres tall. *Muriceopsis* on the other hand are smaller, less bush-like, with side branches arranged more randomly. Microscopic identification of a specimen is often needed to distinguish species. Sea plumes are common, and can form vast forested areas such as that seen near Limestone Bay and parts of Shoal Bay East (as pictured on page 5).

Common Name – **Sea Whips**

Scientific Name – *Pterogorgia* sp.

Description – The sea whips are easy to distinguish as their skeletons are angular and the polyps protrude from their sides. They are usually quite small although some specimens have been seen over 50 cm in height. Considered occasional but can be common on some reef habitats. Although there are other octocorals known as sea whips these get their name from their long whip-like colony growth form and are visually very different from the *Pterogorgia* that, for the scope of this book, are considered sea whips. These other sea whips have not been reported in Anguillian waters to date.

What is a true soft coral? As mentioned at the top of page 65, in the Caribbean the term 'soft coral' has generally come to mean those corals within the Subclass Octocorallia, particularly the gorgonians. Globally, the term actually refers specifically to one particular Order within Octocorallia, namely Alcyonacea, which is only represented in the Caribbean by a couple of deep-water species. However, in the Indo-Pacific region these soft-bodied corals grow in profusion and can form vast colourful forests of gently swaying in the current. Reasons for the lack of them in the Caribbean remains unclear, but the areas younger geological age is probably why, with the Indo-Pacific generally having ten times the biological diversity across all species. Impressive!

Photo Credit Stuart Wynne

Common Name – **Sea Fans**

Scientific Name – *Gorgonia* sp.

Description – Colonies are a network of branches fused together to form a mesh-like structure, and are usually purple in colour. Sea fans are abundant on most reef areas. Although two species are present in the region (see below), the **Common Sea Fan** (*Gorgonia ventalina*) is by far the most common in Anguillian waters. Differentiating between the two species can be difficult underwater, but is based on which surface of the colonies fan-like skeleton is flattened: the outer surface in the Common Sea Fan (pictured top left), and the inner sides in the **Venus Sea Fan** (*Gorgonia flabellum* – not pictured). The growth patterns of sea fans as a whole are generally distinctive although confusingly some species of octocorals also go by the common name 'sea fan'. These mostly belong to the Genus *Muricea* which is usually considered a member of the sea rod group. Interestingly, sea fans were dried and moulded into bowl shapes to be used by indigenous Anguillian people as sieves.

Sea Fan Detail (Photo Credit S. Wynne)

Sea Fan Polyp Detail (Photo Credit S. Wynne)

Notes on other octocoral groups: Although other octocoral groups exist in the Caribbean, none have yet been confirmed in Anguillian waters and so have not been detailed here. These include the telestaceans (Order Telestacea) and the true soft corals (Order Alcyonacea). There are also other miscellaneous species including the sea sprays, some very non-fan-like sea fans that belong to the same Genus as the sea sprays, and other whips, rods and fingers. One example photographed on the wreck of the M.V. Sarah, is known as the **Colourful Sea Rod** (*Diodogorgia nodulifera* - pictured right). Despite its common name, it resembles the telestaceans or sea sprays, and is more closely related taxonomically too.

Photo Credit Stuart Wynne

Coral Disease, Bleaching & Predation

Although other groups of Cnidarians (and indeed other groups of invertebrates) can succumb to disease and predation, the case of the corals is of special relevance to this book because without them (especially the hard corals) there would be a much reduced diversity of life in our surrounding waters and coral reefs would not exist in the form that we know them today.

It is also of special relevance because the number of diseases identified, together with the extent and severity of infection, appears to be increasing. In the 1970s very little was known about coral diseases and prior to 1990 only three had been identified. Even so, they had already been responsible for the suspected demise of vast tracks of reef, and by the end of the 1990s the number of diseases had escalated dramatically. It is not clear whether these diseases have always been present and coral has become more susceptible to them, or whether they are new diseases against which coral has no defence. Whatever the case, corals appear to be becoming increasingly sick with new syndromes documented each year. The overall root cause is very likely to be a reduction in water quality ultimately caused by human activities, although a myriad of factors probably also play a part. Worsening the situation, it seems apparent that corals have reduced their capacity to recover from mortality events. Today, if a colony dies, it is less likely to be replaced by a newly recruiting one. This has a cumulative effect that is all contributing to reef degradation as discussed in Chapter 6 (page 220-230).

Yellow Blotch Disease – This is by far the most common disease presently seen in Anguilla. It seems primarily to affect the corals of the *Orbicella* species complex, with the majority of remaining colonies exhibiting the disease. It is especially apparent in Shoal Bay East (see image below). These Boulder Star Coral species were reportedly far more abundant in recent decades than they are today and this disease is suspected to be the main cause of this decline. The cause of the disease is not clear although reportedly infected corallites have crystalline materials growing inside them. Boulder Brain Coral has also been observed with this disease which starts out as a yellow blotch that slowly expands outwards leaving dead tissue in its centre.

White Band Disease – Historically this disease is documented to be the main cause of the initial coral decline that reportedly occurred in the 1970s. It is thought that this disease, combined with hurricane damage and lack of subsequent recovery, is responsible for the loss of almost all Anguilla's Elk Horn and Stag Horn Coral stands that formed extensive reef systems in the surrounding waters. Only the remnants of these reefs remain now, as they slowly erode away. Some recovery has been seen, although it is by no means substantial. Those on the more exposed south side of the island have since been smashed by storms and piled up along the coast. Little recovery here has been noted. The cause of this disease is not known although a host of microorganisms are usually found at the site of infection. The disease usually starts at the base of a colony and spreads upward, leaving behind the white colonial skeleton as it advances.

Cyanobacteria Infections – The other infections that have been noted in Anguilla from time to time appear to be predominantly caused by cyanobacteria (page 28). Most of the time these plant-like bacteria only cause damage by overgrowing and smothering a colony, a problem exacerbated by nutrient-rich waters. However, coral sicknesses such as Black Band Disease and Red Band Disease have also been attributed to these micro-organisms. Both these diseases have a 'mat' of cyanobacteria that forms a band across the colony, slowly advancing and killing the coral tissue. Infections usually start in a surface depression, or close to a previous injury, and can rapidly spread. The example pictures (over page) shows Black Band disease on Massive Starlet Coral (page 57) and an unidentified cyanobacteria syndrome on *Orbicella* sp., the two species that seem to be most commonly affected.

Coral Bleaching – Coral bleaching results from the loss of symbiotic algae (zooxanthellae) which are expelled if the coral undergoes an elevated and extended period of stress, caused by factors such as raised water temperature and increased levels of pollutants. Although the colony can recover, if it

Orbicella sp. infected with Yellow Blotch disease

Two types of cyanobacteria infection (Photo Credit Stuart Wynne)

remains bleached for too long it essentially starves because it relies on the algae to substitute its nutritional needs through photosynthesis. These zooxanthellae contain pigments and as such are responsible for much of the coral's colour, thus the visual effect of bleaching is that the colony turns white (see image on page 222). Although bleaching has been noted for over a hundred years, mass bleaching events effecting vast areas have only been observed since 1979, with seven major episodes since. The most severe of these were in 1998 and 2005 and great concern exists over their potential to increase in frequency. So much so that the National Oceanic and Atmospheric Administration operate a satellite bleaching alert which is an "automated coral bleaching e-mail alert system designed to monitor the status of thermal stress conducive to coral bleaching via the use of [a] global satellite [system]".

Soft Coral Diseases – Black-Band and Red-Band Disease do not only affect the hard corals, but can also infect gorgonians and other soft corals. Both start at the base of the colony or at sites with an injury, and form a band that radiates progressively across the colony leaving behind it dead tissue. Red-Band Disease is reportedly only observed on sea fans, where as Black-Band Disease can also be found on certain sea plumes. Another disease that primarily affects sea fans is called *Aspergillosis* and is caused by a soil fungus. The infection causes irregular patches to form that lack living tissue, eventually becoming holes surrounded by darker regions. Sea fans can also be seen with tumour-like growths on their surfaces which may result in tissue death and skeletal erosion.

Overgrowth – Both hard corals and gorgonians can be overgrown by other organisms and thus become smothered. Although this usually happens only if a colony has been weakened or killed by disease, as corals are usually able to defend themselves from these invasions (such defences also inhibit corals from growing too close to one another). Fire coral is the classic example of this type of overgrowth which can be commonly seen on most reef environments. More aggressive types of overgrowth can occur, a process

which in essence is the same as the cyanobacteria infections, although the difference is that cyanobacteria infections usually spread in a wave across a colony leaving behind the coral skeleton. In the types of overgrowth covered here the organism usually grows over the top of a colony and successfully establishes itself. Sponges are capable of doing this, as are White Encrusting Zoanthids (page 50). Fortunately, it is rare to see algae overgrowing corals as chemical processes inhibit this. Algae however do commonly outcompete juvenile corals, an increasing problem for the future of coral reefs - as discussed in Chapter 6 (page 220-230).

Fish and Invertebrate Predation – Corals, even those with hard skeletons, are preyed upon by a number of creatures. Fireworms (page 72) for example are an important coral predator and prey on both hard and soft varieties, although reportedly they do have a preference for branching corals and those weakened by disease. Interestingly, damage caused can look very similar to that caused by diseases like white band disease and other white syndromes. It can be differentiated however as fireworm damage originates from the tips of branches whereas disease usually originates from the base. Another important predator, this time solely of the soft corals, are the cyphoma molluscs, most notably the Flamingo Tongue (*Cyphoma gibbosum* – page 89).

These snails typically graze on sea fans, although they will prey on any type of gorgonian, and can cause considerable damage. They are often seen feeding in small groups on a single colony, and will probably be present on most colonies in the vicinity. Other snails also prey on soft corals but they are generally seen closer to the colony base and so are harder to notice, and usually cause only limited damage.

Fish also prey on corals, with butterflyfish (page 135-137) having adapted mouthparts designed specifically to pluck polyps from their corallites. These fish feed so specifically on hard coral polyps that their presence and density is often used as an indicator of reef health. Surgeonfish and parrotfish, although primarily vegetarian, may also graze occasionally/inadvertently on coral polyps. Parrotfish do so quite aggressively and take sizeable bites out of living colonies, often focusing on the same area over a prolonged period causing quite noticeable damage to the colony. The reason for their focused attention is not understood (it may simply be easier for them to take bites this way), but it is thought that their rocky meal may aid digestive health while some nutritional benefit is still gained from the zooxanthellae present in the coral tissue.

3.3 Phylum – Ctenophora: Comb Jellies

Comb Jellies are a small Phylum of transparent, free-floating creatures that are often mistaken for jellyfish. Taxonomically they used to be grouped with jellyfish and the other cnidarians and referred to as coelenterates, although today this is recognised as invalid and obsolete.

Unlike the cnidarians, comb jellies do not have stinging nematocysts. Tentacles are also absent in most species, although a few posses a single pair. The Phylum's most distinctive feature is eight rows of cilia (hair-like structures) called combs which beat and provide the animals with a method of propulsion. This beating motion can produce an iridescent effect which is sometimes mistaken for bioluminescence, although some species are able to produce this effect, which is easily observable at night if they are disturbed.

Divers and snorkellers generally observe comb jellies just beneath the surface where they prey upon small planktonic animals. Sometimes comb jellies will appear in such large numbers that the water seems murky – this usually happens around May/June. Of the species present in Anguilla the most common is the **Sea Walnut** (*Mnemiopsis mccradyi*) which, as its name suggests, resembles a transparent walnut (inset above).

3.4 Phylum – Platyhelminthes: Flatworms

Flatworms, as their name suggests, are flat, worm-like organisms with an oval body that ranges from 2 to 12 cm in length. Some species of flat worm are brightly coloured and may resemble nudibranchs (page 91), although their flat bodies quite easily distinguish them on close examination. Flatworms are not uncommon, but they are rarely sighted by divers because of their highly cryptic habits – living under rocks and in reef crevices. They also often blend extremely well with their background. Interestingly, some species live in the open-water, free-swimming with undulating movements of their entire body

Many species of flatworm have a bad reputation as parasites (tapeworms, flukes etc) although most marine flatworms seen by divers belong to the Class Turbellaria and are free-living and non-parasitic. They are, however, still indirectly linked to their parasitic evolutionary path, being carnivorous scavengers. Flatworms are taxonomically significant because they are considered to be the most primitive creatures possessing a definite front end (with antennae housing rudimentary sense organs) and rear end, also having symmetrical right and left sides. Having said this, only primitive beginnings of specialised organs are present and one opening, situated centrally on the underside, serves as both mouth and anus. Nitrogenous waste however is excreted through surface cells directly into their surroundings. An interesting feature is that many flatworms can regenerate themselves if a part of their body is cut off, sometimes even doing so from the severed part and essentially forming a clone.

Common Name – **Leopard Flatworm**

Scientific Name – *Pseudoceros pardalis*

Description – Purple-brown with orange ocellated spots and a darker margin with small white freckles. This species was confirmed in Anguilla after one was seen swimming close to the surface, although it normally lives a highly cryptic lifestyle under rocks or in reef recesses. Maximum size c.5 cm.

Common Name – **Splendid Flatworm**

Scientific Name – *Pseudoceros splendidus*

Description – Dark blue to black body with orange to yellow undulating margin. As with other flatworms, this species has a highly cryptic lifestyle living under rocks or in reef recesses, and for this reason actual abundances are not known. Maximum size c.5 cm.

3.5 Phylum – Annelida: Segmented Worms

Some species of segmented worms are arguably the most important group of animals on the planet. The earthworms for example, found in many terrestrial environments around the globe, are responsible for circulating the soil and distributing nutrients in such an efficient manner that if they were to disappear suddenly, the floral face of the earth would change beyond recognition. Marine members of this group, characterised by a body made up of repetitive segments, are often overlooked and may be considered less important but are nonetheless fascinating.

Class – Polychaeta: Polychaetes

Polychaetes are characterised by a segmented body with each segment bearing fleshy protrusions housing sensory hairs or bristles. These bristles vary in structure and function, from defence to locomotion to anchorage. Whatever the case, polychaetes are probably one of the most common creatures in the oceans although they are often overlooked because many species live within sediment or burrow into rock. Their sizes vary greatly from the tiny Sponge Worms (*Haplosyllis* sp.), to the Eunices (*Eunice* sp.) which can sometimes grow over a metre in length.

Order – Amphinomida: Fireworms

Fireworms are free-living worms whose sensory hairs have evolved into sharp detachable bristles. When disturbed, fireworms flare these bristles which can easily detach and penetrate a diver's skin, releasing a relatively powerful neurotoxin that can cause pain to the immediate area for a couple of hours. The segmented body, which can grow up to 20 cm long, is usually flattened with gill filaments on each segment that allow respiration. These worms are somewhat reclusive and usually hide under rocks during the day, being a more common sight at night. Sometimes, however, they are seen during the day feeding on soft corals and are considered a significant coral predator (page 70). This feeding behaviour can be seen in the lower right image.

Of the species known in the Caribbean, the most common in Anguillian waters is the **Bearded Fireworm** (*Hermodice carunculata*), the head of which has a large beard-like appendage, with the body segments bearing white tufts of bristles and branched gill filaments.

Order – Sabellida: Tubeworms

Tubeworms consist of two main groups, the feather duster worms and the calcareous tubeworms. Both groups live in tubes but the main difference is that feather duster worms live in a soft tube constructed of sand glued together by a substance secreted from the worm, whereas calcareous tubeworms, as their name suggests, secrete a hard calcareous tube in which they live. Both types have two modified head appendages called radioles that are used to both respire and filter feed for plankton. They are very sensitive to movement, and retract these appendages with the slightest disturbance. The calcareous tubeworms have a hardened structure, known as the operculum, which covers the opening when retracted. Both types are very common in Anguillian waters. Almost any object that finds itself in the ocean will become covered in a large number of other tiny tubeworms that secrete a calcium carbonate tube similar to that of Sea Frost (page 74). Identification of the different species is a highly specialised field and beyond the scope of this book, but the variety present can be observed by looking for differences in the design of the various tubes.

Common Name - **Magnificent Feather Duster**

Scientific Name – *Sabellastarte magnifica*

Description – The largest feather duster worm in Anguillian waters with a crown up to 15 cm in diameter. Unlike other feather duster worms, the radioles are arranged on top of each other rather than side-by-side. Their colours vary from shades of brown to red to tan or even white. Considered uncommon.

Common Name – **Social Feather Duster**

Scientific Name – *Bispira brunnea*

Description – Small crown size of approximately 2 cm, with paired radioles branched separately, the tubes of which are usually exposed. Grow in clusters, and vary in colour from purple-white (as in pictured specimen) to orange-white, the latter being more common in Anguillian waters. Although they will retract into their burrows if closely approached it takes much more disturbance for this to happen than with Solitary Feather Dusters (next). Common.

Common Name – **Solitary Feather Dusters**

Scientific Name – Various (eight Genera)

Description – Species are difficult to identify and usually require microscopic identification by an expert to determine exact classification. All are similar in appearance to the social feather duster, but as their common name suggests, they live a solitary existence. Can be found on reef habitat, amongst seagrass beds or even on sandy bottoms. Although considered abundant, they are often overlooked because they retract quickly into their burrows once disturbed and are generally cryptic in nature. The pictured specimens are the **Split-Crown Feather Duster** (*Anamobaea orstedii* - top), photographed close to Anguillita, and the **Yellow Fan Worm** (*Notaulax occidentalis* - bottom), photographed in Little Bay. The former species can be distinguished by a longitudinal split that divides the crown.

Common Name – **Christmas Tree Worm**

Scientific Name – *Spirobranchus giganteus*

Description – The crowns consist of spiralled, closely paired radioles, each resembling a Christmas tree. These extend from a single aperture that has a hard, trapdoor-like operculum which snaps shut when the worm retracts its appendages. This species is most commonly seen filter-feeding out of the burrows it has carved into coral heads. Abundant.

Common Name – **Sea Frost**

Scientific Name – *Filograna huxleyi*

Description – Grows in colonies of tangled tiny white tubes that can form small clumps. This species usually prefers shady areas and, although considered common, is usually overlooked. Similar in appearance to other species of small calcareous tubeworms that are often seen singly or in small groups encrusted to hard surfaces or sometimes even seagrass fronds.

Order – Terebellidae: Spaghetti Worms

Spaghetti worms are usually overlooked because they hide under rocks or crevices in tubes constructed of fine sand and mud. Their tentacles extend from these hiding places and are only retracted when disturbed or when food has been captured. Identification of different species often requires laboratory examination, although two groups can be broadly identified as detailed following.

Common Name – **Medusa Worm**

Scientific Name – *Loimia* sp.

Description – Translucent tentacles with white spots and rings extend from tube and spread over sand. These species are more common on reef areas or adjacent patches of sand and rubble. They usually remain inside their burrow but individuals occasionally venture out and browse the surrounding habitat. Occasional.

Common Name – **Spaghetti Worm**

Scientific Name – *Eupolymnia* sp.

Description – Body is completely hidden inside a soft tube, out of which extend transparent to white tentacles. These species most commonly live among seagrasses, although because the tube is well hidden they are easily overlooked. Abundant.

Notes on other worm species: There are a great many other types of creatures living in the marine environment which can loosely be described as 'worms'. These include tiny **sponge worms** (*Haplosyllis* sp.), **eunice worms** (*Eunice* sp.) which can be over a metre in length, and **ribbon worms** (Phylum Rhynchocoela). The majority of these are elusive and rarely seen, although chances are greatly increased while diving at night. Some types of worm are so active at night they can actually be quite a menace to divers when they swarm around torches and other light sources.

3.6 Phylum – Arthropoda: Arthropods

Arthropods are the largest Phylum in the animal Kingdom, and include the enormous group known collectively as insects. It is interesting that insects, arguably the most successful Class of animals on land, have not managed to colonise the ocean where crustaceans (see next section) appear to have stolen their crown, and are often referred to as insects of the sea. Currently, taxonomists are disputing whether arthropods should be classified as many different phyla, with the term arthropod being used as a Superphylum heading instead. For the purpose of this book it was deemed appropriate to use the traditional classification.

Arthropods are characterised by an often complex exoskeleton, usually consisting of jointed legs, head, thorax and abdomen. This is achieved by the layering of a tough material, usually chitin, into plates which are connected by flexible protein. The main function of this skeleton, aside from protection, is to avoid desiccation but one main drawback is that it is non-flexible and thus inhibits growth. To grow, arthropods have to completely moult the skeleton leaving them vulnerable to predation. At this time they are soft-bodied and grow rapidly before the surface layer hardens again.

Class – Crustacea: Crustaceans

Crustaceans are the most common arthropods to be found in the ocean and are usually distinguished by a pair of antennae, a series of jointed limbs and a segmented abdomen. This design may differ slightly in some groups (for example the isopods – page 83) or be cryptically disguised (for example the barnacles – page 83), but the basic concept remains the same.

Order – Decapoda: Shrimps, Lobsters, Crabs

The decapods are the largest Order of crustaceans and are characterised by having a fused head and thorax, known as a carapace, which possess five pairs of jointed legs. There are also often other appendages connected to the abdomen but these are usually used for swimming or reproductive strategies and not for walking. The decapods include all shrimps, lobsters and crabs and are one of the most fascinating groups of creatures that live in the sea - from the pelagic swimming crabs to the migrating lobsters that can travel hundreds of miles.

Although there are probably dozens of species of decapods in Anguillian waters this section just contains those most likely to be seen by divers and snorkellers in coastal regions. For example, a species confirmed in 2016, known as the **Caribbean Red Lobster** (*Metanephrops bingham*), is a clawed species and relative of the Dublin Bay Prawn, but being a deep-water inhabitant is unlikely to be seen unless being hauled in a lobster pot.

Common Name – **Banded Coral Shrimp**

Scientific Name – *Stenopus hispidus*

Description – Has a red and white banded body with long, white antennae that can often be seen protruding from under rocks and overhangs. Legs and some body parts are sometimes translucent. In relation to their body size, the third pair of legs possess very large claws that can regenerate if broken. Acts as a cleaner shrimp, picking off parasites from larger creatures. Body length rarely exceeds 5 cm. Common.

Common Name – **Pederson Cleaner Shrimp**

Scientific Name – *Periclimenes pedersoni*

Description – Translucent body with vivid purple colourations on legs and small front claws. Lives in association with a number of anemones (page 45-48) and, as its common name suggests, acts as a cleaner shrimp, picking off parasites from larger creatures. Body length rarely exceeds 2 cm. Common.

Common Name – **Spotted Cleaner Shrimp**

Scientific Name – *Periclimenes yucatanicus*

Description – Transparent body with four pinkish-white body spots, purple and white striped legs, and banded red and white claws. Lives in association with a number of anemones (page 45-48) and, as its common name suggests, acts as a cleaner shrimp, picking off parasites from larger creatures. Body length rarely exceeds 2 cm. Common.

Notes on other shrimp species: Although many other types of shrimp probably live in Anguilla's coastal zone, being cryptic and sometimes only active at night, they are not commonly sighted. Their small size and often transparent body also confounds this. One example is the **Red Night Shrimp** (*Cinetorhynehus manningi*) which is rarely seen during the day, but at night can be sighted in abundance with the aid of a torch. Snapping shrimps are also elusive (*Alpheus* sp.), spending most of their lives hidden away in the reef structure. Their presence is confirmed by the almost constant background crackling that can be heard underwater – a noise made when they snap their claws together, sometimes to stun prey. Other shrimps are often overlooked because they live a camouflaged existence in association with anemones, echinoderms or corals (as with Pederson & Spotted Cleaner Shrimps above), for example the **Sun Anemone Shrimp** (*Periclimenes rathbunae*) and **Squat Anemone Shrimp** (*Thor amboinensis*).

Common Name – **Caribbean Spiny Lobster**

Scientific Name – *Panulirus argus*

Description – Carapace is patchy red and tan colours, while the abdomen is brown with a couple of white ocellated spots down each side. The legs have white vertical stripes, and its antennae are long and cylindrical with a rough surface. It has two very sharp horns over the eyes and lacks the claws for which traditional lobsters are famous. It usually hides in protective recesses during the day but its antennae often poke out and give away its position. This species is very popular in tourist restaurants and was once an abundant resource, but probable overfishing means it is now only an occasional sight while diving although it may still be common in certain areas. It can grow over half a metre in length, with the legal minimum size for fishing in Anguilla being 92 mm carapace length. It is illegal to catch females with eggs.

Photo Credit Lippold Hakken

Common Name – **Spotted Spiny Lobster**

Scientific Name – *Panulirus guttatus*

Description – The body is usually dark brown to red-purple and covered in white spots. This species is known in Anguilla as the Crayfish, although it is not one at all, merely being a clawless lobster. It hides deep in the reef by day and so is rarely seen. It is heavily fished in Anguilla as it is a popular and lucrative dish in local tourist restaurants. Can grow to over 30 cm in length but is usually much smaller. It is illegal to catch females with eggs. Currently considered common, although this status is very difficult to assess due to its cryptic nature and heavily fished status.

Photo Credit Stuart Wynne

Common Name – **Copper Lobster**

Scientific Name – *Palinurellus gundlachi*

Description – Orange to red with no distinctive markings. This species lacks the conspicuous spines of other spiny lobsters but is somewhat hairy with a slightly flattened body. Its front pair of walking legs are almost claw-like and their larger size indicates they play an important role as a foraging tool. This species can grow up to 20 cm in length and is considered uncommon, although actual abundance is difficult to estimate because it lives almost exclusively within the reef structure. The pictured specimen is of an empty exoskeleton found in Shoal Bay East.

Photo Credit Stuart Wynne

Common Name – **Spanish Lobster**

Scientific Name – *Scyllarides aequinoctialis*

Description – Known generically as a Slipper Lobster and locally as the Sea Louse, the body of this species is relatively smooth and generally an even brown colour. It hides in the reef during the day and so is usually only seen while diving or snorkelling at night, when it creeps over the reef slowly scavenging for food. Unlike other lobster species it can easily be picked up, apparently content that its very thick and heavy exoskeleton will keep it safe. May sometimes be found on restaurant menus, but is unreliable to catch so it is not targeted by fishers. Can grow over 25 cm in length and is considered occasional.

Common Name – **Sculptured Slipper Lobster**

Scientific Name – *Parribacus antarcticus*

Description – Carapace is mottled brown to tan with a rough texture and bordered by numerous spines and stiff bristles. Eyes are raised black domes bordered in purple that give this species a very elegant appearance. Can be found on most reef environments where it hides in protective recesses during the day and forages out in the open at night. Can grow up to 18 cm in length and is considered uncommon to rare.

Common Name – **Batwing Coral Crab**

Scientific Name – *Carpilius corallinus*

Description – Has a smooth carapace reddish orange in colour with white spots. A darker area is usually present that looks similar to the shape of a bat with open wings. Hides in protective recesses during the day, venturing out only at night, so is not commonly sighted. May sometimes be found on restaurant menus, but is unreliable to catch so it is not targeted by fishers. Its carapace can grow 15 cm in width. Common.

Common Name – **Hermit Crabs**

Description – Marine hermit crabs are common in Anguillian waters although, as with many other crustacean species, are often overlooked because they are cryptic and quickly retreat into their shells when approached. They use a variety of mollusc shells as their home, discarding their current choice for a larger or otherwise more favourable one as they grow. Some species can be quite large, such as the **Giant Hermit** (*Petrochirus diogenes* - top) which may grow up to 30 cm in length, while others can be less than a centimetre in size. Many species look similar and are hard to identify, some needing laboratory examination to be sure of correct species identification. The **White Speckled Hermit** (*Paguristes punticeps* - bottom left) is probably the most common of these smaller species. Terrestrial species, most notably the **Caribbean Hermit Crab** (*Coenobita clypeatus* - bottom right) are numerous all over the island. Known locally as Soldier Crabs, they are especially obvious when returning to the sea to cast their eggs into the water around September each year. Aside from this habit, they live an entirely land-based way of life.

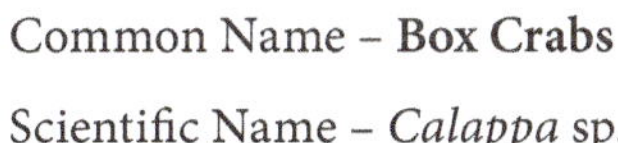

Common Name – **Box Crabs**

Scientific Name – *Calappa* sp.

Description – Stout, box-like crabs of varying colours whose claws are barely visible once folded perfectly into indents in their carapace. Most species spend much of their lives hiding in recesses or buried under the sand with only their eyes showing. The pictured species is the **Flame Box Crab** (*Calappa flammea*) seen at Fish Hole Pond on Scrub Island, although the **Rough Box Crab** (*Calappa gallus*) is probably the most common species in Anguillian waters. As box crabs spend most of their time partially buried on sandy bottoms, they are normally overlooked.

Common Name – **Swimming Crabs**

Scientific Name – *Portunus* sp. & *Callinectes* sp.

Description – Bodies are usually flattened with sharp elongated claws and paddle-like back legs which they use to swim through the water. Although identification to species level is often difficult the pictured specimen is probably a member of the Genus *Callinectes* (**blue swimming crabs**) which are usually larger than other swimming crabs. In actual fact, many of the most commonly sighted swimming crabs in Anguillian waters are white and only a centimetre or two across, so they are often overlooked as they quickly scuttle across the bottom before burying themselves under the sand.

Common Name – **Nimble Spray Crab**

Scientific Name – *Percnon gibbesi*

Description – Has a flattened carapace and legs that allow it to squeeze into very thin gaps in the reef structure and under rocks. This habit means they are rarely seen by snorkellers, but inquisitive divers may see their yellow banded legs when peering into crevices. Its cryptic nature make abundance estimates difficult, but it is thought to be common with a carapace width not much larger than c.5 cm. Close relatives of the Nimble Spray Crab, other **flat rock crabs** from the Genus *Pachygrapsus* live between rocks in the splash zone and only spend occasional periods submerged. *Pachygrapsus* sp. can be observed on almost any rocky area around Anguilla's coastline.

Common Name – **Yellowline Arrow Crab**

Scientific Name – *Stenorhynchus seticornis*

Description – This curious creature does not look much like a crab at first glance, more resembling a spider. It has a golden brown triangular body with long legs, tiny claws, and a covering of fine yellow lines. Its carapace is only a centimetre or two in width, which combined with its habit of hiding in recesses, means it is often overlooked. Common to occasional.

Common Name – **Clinging Crabs**

Scientific Name – *Mithrax* sp.

Description – Most clinging crabs are tiny and vary in colour within species, making identification very difficult. The largest and probably most common species in Anguillian water however is called the **Channel Clinging Crab** (*Mithrax spinosissimus* - top) and is easily identifiable for the antithesis of these reasons. It is reddish-brown in colour with large, smooth, purplish claws and a carapace that grows to a width of 20 cm. It is the largest Caribbean reef crab and a popular seafood, although not usually locally available in Anguilla. *M. spinosissimus* goes by many other names including the Reef Spider Crab and King Crab. Of the other clinging crab species present in Anguillian waters the **Red-Ridged Clinging Crab** (*Mithrax forceps* - bottom) is probably the most likely to be seen.

Common Name – **Neck Crabs, Decorator Crabs & Sponge Crabs**

Description – These crabs, belonging to a variety of Genera, have the fascinating habit of attaching other living organisms to their bodies as a means of camouflage. The habit is so effective that even common species are hardly ever seen and identification is extremely difficult, with attached organisms usually remaining alive and even reproducing. The individual pictured to the right is one such unidentified species. Notes on other crab species can be found overleaf, which includes an example of a sponge crab, probably from the Genus *Dromia*.

Notes on other crab species: The crab community in Anguillian waters is much more diverse than represented here with probably dozens more species present. Most of these species however are very small and so often overlooked, usually living a camouflaged existence in close association with anemones, corals and sponges – for example the **Banded Clinging Crab** (*Mithrax cinctimanus*) and **sponge crabs** (*Dromia* sp.). Also small in size, the porcelain crabs (not pictured) look similar to true crabs, but are taxonomically more closely related to hermit crabs. They have long antennae and a reduced number of walking legs, the fourth pair of which have become small spike-like appendages close to the abdomen. There are also land crabs, other than the Soldier Crab (page 79), that deserve a mention here. The **Atlantic Ghost Crab** (*Ocypode quadrata*) is common on many beaches where it lives in burrows and forages amongst the sand, whereas the **White Land Crab** (*Cardisoma guanhumi*), lives in relatively elaborate underground tunnels and chambers close to salt ponds. This latter species can become quite sizeable and is a popular food source in many parts of the Caribbean, although it is not widely eaten in Anguilla probably because it is not very common.

Red Eye Sponge Crab (*Dromia* sp.)

Banded Clinging Crab (*Mithrax cinctimanus*)

Atlantic Ghost Crab (*Ocypode quadrata*)

White Land Crab (*Cardisoma guanhumi*)

Order – Stomatapoda: Mantis Shrimps

Mantis shrimps are, as their name suggests, shrimps that resemble a Praying Mantis because of their front appendage similarities. They usually live in burrows, although sometimes occupy crevices in reef habitats. They have arguably the most complex set of compound eyes in the animal Kingdom which are housed on stalks and move independently. It has even been suggested that their sight is so good they can recognise different coral types. Three pairs of walking legs differentiate them from the decapods (previous). Their mantis-like claws are extremely powerful and sharp, being used for catching prey and are even able to slice into an unsuspecting finger if it is placed within range. Species can be difficult to identify, changing colour to match their background. The pictured specimen (left) is probably the **Ciliated False Mantis** (*Pseudosquilla ciliate*), with its front claws folded back into special slots, and so are not clearly visible.

Order – Isopoda: Isopods

Marine isopods, known more specifically as cymothoid isopods, look like parasites but actually do not directly harm their hosts as they are merely catching a ride in much the same way as sharksuckers do (page 197). Isopods favour certain species of fish, for example chromis (page 143), creolefish (page 152) and squirrelfish (page 167-168), usually attaching singly to each cheek. They start out life as free-swimming males but once on their host they lose the ability to swim. They attach with several pairs of hook-like appendages and scavenge for small particles of food that pass by them. The ultimate goal of the isopod, as with all other creatures, is to find a mate. Males have the interesting ability to transform into females, which increases their chances of reproductive success; as once they attach to their host they stay for life. If a male settles and a female isn't present, he accelerates his growth and turns into one, ready for the next arriving male. If a female is already in place when the male arrives then he mates with her and when she dies he turns into a female himself. There are at least a dozen different species belonging to two Genera (*Anilocra* & *Renocila*) although identification is extremely difficult and usually requires microscopic examination.

Brown Chromis with Isopod (Photo Credit S. Wynne)

Order – Mysidacea: Mysid Shrimps

Mysid shrimp are tiny (<3 mm) shrimp-like creatures that usually appear as small clouds made up of white or almost transparent specks. They can easily be mistaken for fish larvae, and are often seen swarming around Long-Spine Sea Urchins (page 99). Three main Atlantic species are known, all belonging to the Genus *Mysidium* (not pictured), but their small size means microscopic examination is required to distinguish them.

Order – Thoracica: Barnacles

Barnacles are almost colonial crustaceans that fix themselves permanently to a variety of usually solid surfaces. They are very often found on floating material in the water and thus can be a useful guide to assess how long something has been left afloat. They are famously described as shrimp-like animals which stand on their heads filter feeding with their feet, although at first glance this analogy is difficult to understand. In fact, superficially barnacles don't look like crustaceans at all because they are surrounded by a series of hard protective plates. Their crustacean history is however revealed during their planktonic phase, when they closely resemble a larval crab. The Order Thoracica is often considered a Superorder containing Pedunculata, the gooseneck barnacles, and Sessilia, the 'classic' barnacle. For the purpose of this book however they have been grouped together as one. **Sessile barnacles** (*Balnus* sp. & *Megabalanus* sp. - not pictured) are cone-shaped and permanently fixed to objects at the base of their shells. The outer shell is usually composed of six to eight plates, two of which are beak-like and articulated, allowing the animal to reach out and filter-feed. Identification of the different species is usually by examination of the barnacle itself, not its shell, and so is not possible for the casual observer. They are often found growing on mature turtles, large gastropods or on rocks.

Common Name – **Smooth Goose-Neck Barnacle**

Scientific Name – *Lepas anatifera*

Description – The animal's body is encased in five flattened white shell plates that are articulated, allowing the animal to reach out and filter-feed. The shell is attached to a brownish flexible stalk that may reach up to 3 cm in length when fully extended. The pictured individual is attached to a boat mooring line. Common.

3.7 Phylum – Ectoprocta: Bryozoans

Bryozoans are very small colonial animals that in many ways resemble corals. Each animal, known as a zooid, has polyp-like tentacles around the mouth. These filter feeding tentacles extend from a skeletal chamber that the creature secretes around itself. The chambers are not usually much larger than 1 mm in diameter and can vary in shape from oval to rectangular 'boxes'. One fundamental difference from corals is that bryozoans have a complete digestive system, including an anus that lies outside the ring of tentacles. Corals on the other hand are far simpler creatures with a combined mouth and anus, and only a primitive digestive tract.

The colonies that these zooids form can also vary in shape from flat, encrusting varieties (inset above) to those that resemble small sea fans or even seaweed. The chamber is calcareous although chemical composition varies, meaning that some colonies are flexible while others are rigid and somewhat fragile. Identifying a growth as a bryozoan is achieved by looking for this zooid arrangement, with species identification usually based on its shape.

3.8 Phylum – Mollusca: Molluscs

Molluscs are soft-bodied creatures, classically with a hard external shell although there are actually a number of variations on this plan. The shells are made from calcium carbonate and are slowly secreted over time by a specialised layer of tissue known as the mantle.

Class – Gastropoda: Gastropods

Gastropods are the largest group within the molluscs and contain creatures with both the classic, coiled snail shell design, and the shell-less, slug-like design. Those with shells have a slightly complex taxonomy and so are presented below as one large morphological group rather than being split into Orders, whereas the shell-less snails are grouped more classically. Gastropods are characterised by a process called torsion, in which the internal organs of the animal rotates 180 degrees during embryonic development placing the anus almost directly above the head. This process should not be confused with the familiar coiled shape which many of the shelled gastropods have.

Gastropods are the second largest Class of animals in the world, only being beaten by the insects. Of the 80,000 known living species, two thirds live in the marine environment, from the abyssal depths to the pounding surf of rocky shores.

Gastropods with shells

Gastropods (derived from the Latin words for stomach and foot) are characterised by the coiled snail shell that most members possess, although the limpets (page 89) and some other Families only possess a coiled shape in their larval stage, becoming a simple conical structure later in development.

Many more species than those listed here can be found in Anguillian waters, but those that have been included comprise the more common or conspicuous ones. Other species may be common but because of their cryptic habits are rarely seen until their empty shells wash up on beaches. These latter gastropods have been included in Chapter 5 (page 209-219).

Common Name – **Queen Conch**

Scientific Name – *Lobatus gigas*

Description – The largest member of this Genus in the Caribbean (max c.30 cm), with a short conical spire and blunt spikes. The shell exterior is orange, although it is often covered with algal growth and sediment aiding camouflage, and the inside of the aperture is vivid pink. Their head is mottled grey and they have a claw-like operculum that they use to drag themselves across the sea bed. Juveniles live under the sand until they reach 5 to 10 cm in length, with their spires and spikes being more pronounced. As maturity is reached they slow their growth and begin thickening their shells by adding layers to it internally. This, combined with gradual erosion of the shells outer surface, blunts their overall appearance. The shell aperture begins to develop a pronounced lip and when this lip thickens to approximately 5 mm conch are considered fully mature. Once abundant on seagrass beds and sandy areas Queen Conch are now less common due to overfishing – a widespread occurrence throughout the Caribbean where they are considered a delicacy. The top photograph is of an immature individual, whereas the bottom image illustrates the thickened lip, as well as some very curious eyes looking out. Formerly classified under the Genus *Strombus*.

Common Name – **Milk Conch**

Scientific Name – *Strombus costatus*

Description – Has a thick, whitish shell and aperture with short conical spire and round spikes. The head is green and they have a claw-like operculum used to drag themselves across the sea bed. Prefer seagrass areas and sand flats. Occasional, with a maximum size of c.15 cm.

Notes on other conch species: Although other species of conch will certainly live in Anguillian waters, only a small number of sightings have been made. Of these species **Hawkwing Conch** (*Strombus raninus*) and **Roostertail Conch** (*Strombus gallus*) are the most likely to be seen. Interestingly, the shell of this latter species is often found inhabited by hermit crabs (page 79) in the waters around Prickly Pear Cays.

Common Name – **True Tulip**

Scientific Name – *Fasciolaria tulipa*

Description – Smooth conical shell mottled reddish-orange in colour with thin dark spiral pattern encircling it. The abundance of this species has not been confirmed but, based on the frequency that it washes up on the beach, is probably uncommon. Prefers shallow sand and seagrass areas, especially in sheltered bays. Maximum size c.25 cm.

Common Name – **Atlantic Triton Trumpet**

Scientific Name – *Charonia variegata*

Description – Long pointed spiral shell varying in colour from cream to dark brown. Prefers sandy areas close to reefs where it hides in protective recesses during the day. Its beautiful patterns have led to it becoming rare in many parts of the Caribbean due to over-collecting, however it is unclear whether this is the reason for its uncommon status in Anguilla. Interestingly it reportedly targets sea cucumbers (page 102-103) as a food source. Maximum size c.35 cm.

Common Name – **West Indian Topshell**

Scientific Name – *Cittarium pica*

Description – Known locally as a Whelk, this species forms a flattened spiral with alternate black and white banding patterns. The pattern is often covered with a growth of encrusting calcareous algae and so can be quite obscure. This species is rarely seen by snorkellers as it lives in the surf zone in rocky areas, fixing themselves solidly into rock depressions with a firm suction grip. It is a popular local activity to go 'whelking' where people either brave the surge close to rocks with a mask and snorkel or pick them directly by walking close to the splash zone. Their operculum is a thin, black, semi-transparent flat spiral that can sometimes be found washed up on beaches and piles of broken shells are often found close to their preferred habitat as a remnant of people whelking (Chapter 5 - page 209-219). Common to abundant on rocky shores, growing up to c.15 cm wide.

Common Name – **King Helmet**

Scientific Name – *Cassis tuberosa*

Description – Has a blunt spire and thickened triangular aperture lip with dark stripes. Colourations of cream to reddish-brown make wave patterns over its shell. These beautiful patterns have led it to becoming rare in many parts of the Caribbean due to over-collecting, but it is not known how prevalent this practice is in Anguilla. Although considered uncommon, the actual abundance of this species is difficult to establish because they burrow into the sand during the day and leave only part of the upper shell exposed. During the night they hunt urchins with surprising speed. Maximum size c.15 cm. A less common but very similar species, the **Emperor Helmet** (*C. madagascariensis* - bottom), can grow up to twice this size. It can also be differentiated by fewer shell markings and a less triangular lip around its aperture.

Common Name – **Nerite Snails**

Scientific Name – *Neritia* sp.

Description – These common small molluscs look like juvenile West Indian Topshells (previous), although they belong to a Family of their own. There are a number of species present on Anguilla's rocky shorelines, although identification between them can be quite difficult (pictured specimen is probably *Nerita versicolor*). They are abundant in the intertidal zone (between low and high water marks), although they may be present in the splash zone and also below low water mark. Considered an important grazer, these creatures are also often favoured by hook & line fishers who take advantage of the plentiful resource that is present at their favourite fishing spots. Some species from this Genera are small with highly intricate designs on a smooth-surfaced shell (page 213), and thus appear very different from the more commonly found varieties.

Common Name – **Stocky Cerith**

Scientific Name – *Cerithium litteratum*

Description – Long pointed spire with whitish shell covered in dense blackish blotches. This species is small (max c.2 cm), but can be easily spotted as it commonly aggregates in quite large numbers. Can be found in a wide range of habitats including seagrass beds, under rocks, on rubble areas, or in intertidal pools. Locally abundant.

Common Name – **Periwinkle**

Description – Although these species do not generally live underwater, because they exist in close association with the ocean (in the splash zone) they have been included here. Identification to species level is difficult, but those present are likely the **Prickly Periwinkle** (*Nodilittorina tuberculata*), the **False Prickly Periwinkle** (*Echininus nodulosus*), and the **Beaded Periwinkle** (*Tectarius muricatus*). Their coiled conical shells are usually grey in colour and covered with small bumps. Maximum size c.2 cm. Abundant on almost all rocky shores.

Common Name – **West Indian Star Snail**

Scientific Name – *Lithopoma tectum*

Description – Has a relatively high conical spire and, although light brown in colour, is often overgrown with red coralline algae (page 26). This species is cryptic, often living in seagrasses or under rocks, but it is sometimes seen in shallow pools or on occasion wandering the reef feeding. Considered common. A similar species which also occurs here but usually lacks the red algae is known as the **Astrea Snail** (*Astraea tecta*).

Common Name – **Flamingo Tongue**

Scientific Name – *Cyphoma gibbosum*

Description – Shell is white and almost olive-shaped with a smooth ridge running across its middle. The mantle usually extends all the way around the shell and is cream-white in colour with orange ocellated spots. This intricate pattern tempts people to collect the shells when diving, but once disturbed the mantle retracts and reveals its somewhat plain-looking exterior. This species attaches and feeds on gorgonians and, if in high enough densities, can inflict considerable damage to a colony. Common on soft coral dominated reefs although can be locally abundant. Maximum size c.2.5 cm.

Common Name – **Fingerprint Cyphoma**

Scientific Name – *Cyphoma signatum*

Description – The shell is similar to that of the Flamingo Tongue (previous) although it is noticeably more elongate. The mantle usually extends all the way around the shell and is cream-white in colour with orange fingerprint-like patterns over it. Although considered uncommon to rare this species has been sighted a number of times in parts of Shoal Bay East, where it was attached to and feeding on colonies of Slit-Pore Sea Rods (page 66) causing them considerable damage. Maximum size c.2.5 cm.

Common Name – **Keyhole Limpets**

Scientific Name – Family - Fissurellidae

Description – Relatively flat conical shell with a hole in the apex. The members of this Family are not in fact true limpets (Family Patellidae), but a more ancient group consisting of only distantly related species. Most keyhole limpets in Anguillian waters probably belong to the Genus *Fissurella*, although precise identification is difficult, in part because they are often covered with turf algae and sediment. Keyhole limpets spend most of their lives in the littoral zone of rocky areas where they are considered abundant.

Notes on other shelled mollusc species: There are many other species of molluscs that occur in Anguillian waters and its coastal zone although many are uncommon and this, combined with their cryptic nature, means they are rarely spotted. Evidence of these species is found when their shells wash up on beaches, a selection of which are detailed in Chapter 5 (page 209-219).

Gastropods without shells

Gastropods without shells, sometimes referred to as shell-less snails, do not have a true shell into which they can retreat for protection. Of the five distinct Orders, some do possess a small external shell while others only have a small internal one. Three of these Orders are described below, the remaining two have yet to be reported in Anguillian waters and so have not been included. These latter Orders are the headshield slugs (Cephalaspidea) and the sidegill slugs (Sacoglossa).

Order – Anaspidae: Sea Hares

Sea hares are relatively large soft-bodied gastropods that look like rather elaborate slugs. They have a pair of rolled tentacles that seem to feel the ground directly in front of them, a pair of rolled sensory tentacles called rhinophores that are positioned where one would expect their eyes, and mantle flaps that are attached to their backs and resemble wings. They have a small internal bone, internal gills and feed on algae. Most species are quite drab, usually being greens and/or browns. There is only one species currently confirmed in Anguillian waters, the **Spotted Sea Hare** (*Aplysia dactylomela*). These sea hares are light brown and covered with an irregular pattern of thick, black, ocellated spots.. In a similar manner to squid, if disturbed this species will discharge a harmless purple fluid that distracts potential predators. Can grow up to a maximum size of c.30 cm and prefers to live in grass areas with scattered rocks. Seen quite frequently in the Katouche Bay area, but are generally considered occasional.

Order – Sacoglossa: Sea Slugs

Sea slugs are a diverse Order that, strangely enough, is united by one unique feature – a sac in the alimentary canal that collects discarded teeth! Identifying a creature as a sea slug is usually achieved by observing a pair of rolled sensory tentacles called rhinophores and an often elaborate set of skin ruffles on their back. These ruffles can be the key to species identification and serve to increase the creature's surface area aiding absorptive processes. Most sea slugs have an internal shell, and spend much of their time feeding on algae. Of those species present in Anguilla, the **Lettuce Sea Slug** (*Elysia crispata*) is undoubtedly the most common as it is the only species that has been positively identified. Its colour varies considerably, but is usually green with beautifully ornate skin ruffles. Occasional.

Order – Nudibranchia: Nudibranchs

Nudibranchs can be recognised by their external gills which are located towards the rear of the animal. They are the only true shell-less snails as, unlike the previous groups, they lack either an external or internal shell. They have a pair of upward pointing sensory tentacles called rhinophores and a secondary pair of oral tentacles that, in some species, are small and hard to see. Some have developed skin ruffles similar to those in the sea slugs which increase their oxygen absorption area, while others have fringe-like projections that serve the same purpose. Some species can free-swim while most slither around their habitat in a more slug-like way and, being carnivorous, hunt for food.

Nudibranchs are usually colourful and elegant members of the marine fauna, although their small size means they are often overlooked and sometimes difficult to identify. They can also be easily confused with sea slugs (previous) and flatworms (page 71). There are many species which have been recorded in the Caribbean, although few have yet to be positively identified in Anguilla. In fact, the only confirmed sighting is of the **Slimy Doris** (*Dendrodoris krebsii*) seen close to Anguillita Island. However, of those present it is likely the majority belong to three Genera: *Hypselodoris, Chromodoris & Dendrodoris.* Species may include the **Harlequin Blue Sea Goddess** (*Chromodoris clenchi*) - photograph not taken in Anguilla.

Slimy Doris (*Dendrodoris krebsii*)

Harlequin Blue Sea Goddess (*Chromodoris clenchi*)

Class – Amphineura: Chitons

Chitons (local name – Hardback) can easily be recognised by their oval shape and segmented external 'shell', made of overlapping calcareous plates. They attach themselves firmly to rocks with a muscular foot and graze algae, but their action is so slow it is rarely possible to see them move and many people mistake them for fossils or small rock formations. Chitons are usually found in the intertidal zone but they have been included here because occasionally they are found on reefs. Of the species found in the Caribbean only one has been confirmed in Anguillian waters, the **Fuzzy Chiton** (*Acanthopleura granulata*). This species has brown plates that often become encrusted with algae and a grey girdle with black bands and short, coarse, hair-like spines. They are abundant on most rocky areas and grow to a maximum size of c.9 cm.

Class – Bivalvia: Bivalves

Bivalves are so called because of two hinged shells, or valves, that encase their soft, fleshy body. These valves are usually of equal size and have muscles that can open and close them allowing an appendage, known as the siphon, to be extended. The siphon, which can be quite sizeable, is usually used as a water intake allowing the creature to filter feed. Some species also extend part of their mantle through the valves which possess sensory tentacles that can detect movement. Bivalves are usually identified by their shell shape and surface features although some species can be very similar to one another and so require closer examination. Many members of this group spend their lives buried underneath the sand. As these species are not usually sighted until after death, when their shells wash up on a beach, they have not been included in this chapter. A small selection of potential finds has however been presented in Chapter 5 (page 209-219).

Common Name – **Rough File clam**

Scientific Name – *Lima scabra*

Description – Brilliant red to orange mantle with tentacles protruding from brownish valves patterned with fine radiating ribs. Usually lives in narrow cracks in the reef so is easily overlooked, but if startled can swim in a jerky fashion by rapidly opening and closing its valves. Although rarely sighted this species is considered common and can grow up to 9 cm in diameter. Others of this Genus are present in Anguillian waters, evidenced by their washed up shells, but no live specimens have yet been recorded.

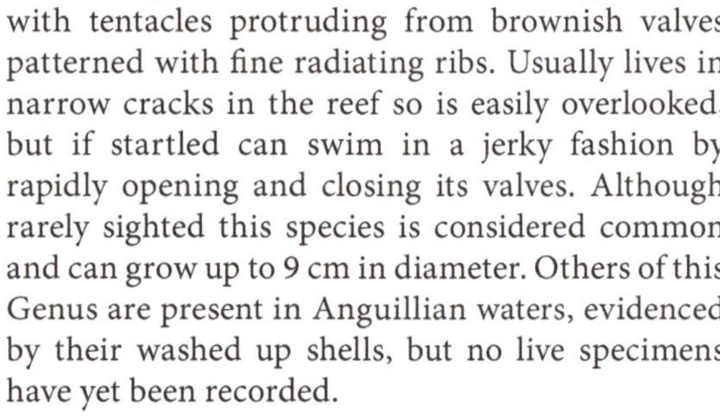

Common Name – **Atlantic Thorny-Oyster**

Scientific Name – *Spondylus americanus*

Description – Valves covered with scattered curved spines which may be several centimetres in length. Shell orange to white. Has a brown mantle with white markings. Is considered common but lives a cryptic existence under overhangs and ledges, and is often disguised by overgrowing algae. Maximum size 12 cm.

Common Name – **Atlantic Wing-Oyster**

Scientific Name – *Pteria colymbus*

Description – Dark brown, with a long wing-like growth that extends along the valve hinge. Is commonly found attached to mooring buoy lines and grows naturally on the stems of soft corals (page 65-68). Common, although is often well camouflaged by an overgrowth of a variety of organisms. Maximum size 7 cm.

Common Name – **Frond Oyster**

Scientific Name – *Dendostrea frons*

Description – Irregular-sized valves that are characterised by a pronounced zig-zag pattern at their opening. Is commonly found attached to mooring buoy lines and grows naturally on the stems of soft corals (page 65-68). Common, although is often well camouflaged by an overgrowth of a variety of organisms. Maximum size c.5 cm.

Common Name – **Amber Penshell**

Scientific Name – *Pinna carnea*

Description – Grey translucent valves that are similar in shape to a partially closed fan. They live half-buried vertically in sandy or muddy bottoms and so are often overlooked. May also be found living in seagrass beds, but are almost impossible to spot. Sometimes they can be found inhabiting narrow crevices within the reef. Common, with a maximum length of 15 cm.

Class – Cephalopoda: Squid and Octopuses

Cephalopods are probably the most intriguing members of the mollusc Family, being highly intelligent and at first glance very different morphologically from the rest of the Class. In fact, some groups of cephalopods are deemed so intelligent that research conducted on them has to follow a similar ethical framework as that applied to higher vertebrates. This essentially confirms their place as the only 'intelligent' invertebrate group.

When their design is simplified it is possible to see the similarities that cephalopods share with other molluscs as they are soft-bodied creatures with an internal bone and a series of tentacles. On closer examination, however, their complexity begins to amaze with a highly complex nervous system controlling an intricate array of cells (chromatophores) which allow them to change body colours extremely rapidly, often in bizarre and multifaceted ways. They also have highly advanced eyes and intricate behavioural patterns, with locomotion powered by a mantle cavity that expels water with a jet-like action.

All cephalopods have cup-like suckers on their tentacles used to catch prey and manoeuvre it towards their bone-crushing, powerful, beak-like jaws. These suction cups have horny rims which in larger species can inflict noticeable damage on prey. Cephalopods vary in size and habitat from the tiny, shallow water and highly toxic Blue Ring Octopus, to deep sea Colossal Squid which are amongst the largest creatures in the sea. These giants are so big they leave huge scars when wrestling with Sperm Whales which prey on them for food.

Order – Teuthoidea: Squid

Squid are distinguished by eight short 'arms' and two longer tentacles attached to the front of their elongate body. They have stabilizing fins on the sides of their bodies that undulate elegantly as they swim, and if startled they zoom backwards powered by a jet-like puff of water from their mantle, leaving behind a cloud of thick 'ink' to distract potential predators. They are often seen swimming in large groups and prey on small fish. Of the Caribbean species two have been confirmed in the coastal waters of Anguilla (see below), along with the **Diamondback Squid** (*Thysanoteuthis rhombus*), a large, deep-water species. This species was confirmed during an exploratory fishing study conducted by the Department of Fisheries and Marine Resources in 2013. It is also thought that the **Longfin Squid** (*Loligo pealei*) is common in Anguillian waters, but its preference for deeper water means it is unlikely to be witnessed by divers or snorkellers.

Common Name – **Caribbean Reef Squid**

Scientific Name – *Sepioteuthis sepioidea*

Description – Thin fins border the entire body, tapering into a point at the end. Colour varies, although during the day is usually brownish lavender with white spots. Maximum size c.30 cm, but is usually a lot smaller than this. Common, often forming huge groups that are frequently sighted over seagrass beds (for example Road Bay).

Common Name – **Arrow Squid**

Scientific Name – *Loligo plei*

Description – Triangular fins border the rear half of elongated body. Usually pinkish-red, with adult males being distinguished by flame-like markings on side of body. Maximum size c.40 cm but is usually a lot smaller than this. Uncommon. Seen in mixed groups with Caribbean Reef Squid (previous).

Order – Octopoda: Octopuses

Octopuses (octopodes is the classically correct plural) have eight arms of equal length and an extremely soft, yet tough-skinned, bell-shaped body. Movement is usually controlled by their highly dextrous tentacles, which they use to crawl across the reef surface, but a powerful jet propulsion system similar to that of the squid can be employed to rapidly escape predators. During this escape they also often expel a cloud of thick 'ink' to confuse their enemy. They can fold their bodies into extremely tight recesses and squeeze through very thin gaps. This ability, combined with highly effective camouflage, means these nocturnal hunters are rarely seen during the day. This camouflage, driven primarily by the chromatophores, is so advanced that even the texture of the skin can be changed to match the surrounding habitat. Becoming active and bolder after dusk when they hunt crustaceans and shelled molluscs, the best chance to see an octopus comes when undertaking a night dive.

Common Name – **Caribbean Reef Octopus**

Scientific Name – *Octopus briareus*

Description – Colourations can change rapidly but seem to favour lighter tones with green and brown mottling. Suckers do not have dark edges (compare Common Octopus next), and skin is relatively smooth. The eyes often have a dark ring around them and the arms are four to six times its body length. Although this species is believed to be common the exact status is unknown as, being highly cryptic with the ability to squeeze through the smallest of gaps, this is very difficult to estimate. Maximum size c.45 cm.

Common Name – **Common Octopus**

Scientific Name – *Octopus vulgaris*

Description – Colourations can change rapidly but seems to favour mottled brownish-tan-red. Suckers have dark edges (compare Caribbean Reef Octopus previous), and skin is usually lumpy and/or reticulated, although this can be varied. The eyes do not have a dark ring around them and the arms are three to four times its body length. Although this species is believed to be common the exact status is unknown as, being highly cryptic with the ability to squeeze through very small gaps, estimations are difficult to make. Maximum size c.45 cm.

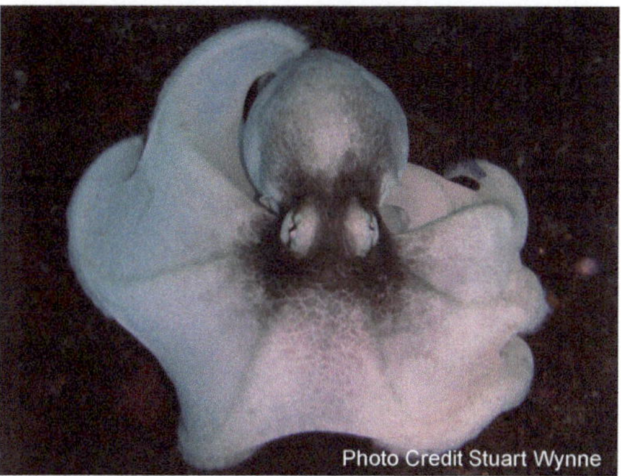

3.9 Phylum – Echinodermata: Echinoderms

Echinoderms are, at first glance, a varied group of creatures. Common to all members is a body arrangement split into five sections of equal size which leads to many groups having a very obvious radial symmetry. All echinoderms have a hard skeleton made up of small calcareous plates called ossicles. This skeleton is visible in most groups, but in the sea cucumbers the ossicles have been reduced to tiny fragments buried in their leathery skin. Another key feature is that most have hundreds of tube feet called podia, which work in unison to provide locomotion and/or to capture and manipulate food. In fact, the movement of these podia are often the only clue as to whether or not an echinoderm is alive.

Class – Crinoidea: Feather Stars

Feather stars, or crinoids as they are often called, are usually overlooked or mistaken for plants. They have a very small, flattened body that is usually tucked away in a reef recess, with five arms which almost immediately branch one or more times. The arms have structures down each side giving a feather-like appearance, which are used to filter-feed as they are swept around in the water. As with sea stars and brittle stars (following), their arms can be regenerated if damaged. Some species of crinoid can swim freely with flapping movements of their arms, while most others are able to walk along the reef surface with jointed legs when not anchored in a crevice. Interestingly, although some species of crinoid are considered abundant in the Caribbean (**Golden Crinoid –** *Davidaster rubiginosa*), sightings have yet to be confirmed in Anguilla.

Class – Asteroidea: Sea Stars

Sea stars, misleadingly known in the past as starfish, usually have five arms of a generally triangular shape, merging at the base of a central disc. The arms have the useful capacity to regenerate themselves if cut off and some species can even re-grow a new body and four new arms from a severed piece, essentially being a cloning process. The anus is situated on the top of the central disc and the mouth underneath, with four rows of podia radiating from it along the underside of the arms. The podia have little suction cups on the end and are used for both locomotion and prey capture. Some species of sea stars have the curious habit of extending their stomachs out through their mouths and consuming their prey externally. Of the sea star species in the Caribbean, only a few are found commonly in Anguilla and the majority of these are nocturnal so rarely seen.

Common Name – **Cushion Sea Star**

Scientific Name – *Oreaster reticulatus*

Description – Colourations vary, although individuals are usually orange to reddish-tan. The legs are short and thick, and the whole creature is covered with knobbly blunt spines. Colorations between these spines form complex, often geometric patterns. Carnivorous, preying on a variety of molluscs. Genetic variants exist that can commonly have six legs, or in rare cases four. Abundant to common on sand flats and seagrass beds (especially juveniles), with a maximum diameter of over 30 cm.

Photos illustrating the diversity of Cushion Stars found around Anguilla. Far left - two different colour morphs; Top middle - rare genetic variation displaying four legs; Top right - genetic variation displaying six legs; Bottom middle - juvenile living cryptically among turtle grass in Merrywing Bay.

Common Name – **Striped Sea Star**

Scientific Name – *Luidia clathrata*

Description – Has an obvious border of white spines and a rough textured upper surface that is usually cream to grey. Arms often have a darker central stripe, although this is obscure in the pictured specimen. Common but nocturnal, living under the sand during the day.

Class – Ophiuroidea: Brittle & Basket Stars

Members of this Class have long, thin arms that widen very little as they reach the central disc, which is a flattened, somewhat rounded and usually smooth structure. Their mouth is located in the middle of the underside of this central disc but interestingly, they lack both intestines and an anus. This group is split into two distinct Orders.

Order – Ophiurida: Brittle stars

Brittle stars are generally small and reclusive creatures living under rocks and in crevices, so are rarely seen by underwater observers. They have a small, smooth central disc usually less than 3 cm in diameter and long fragile-looking arms covered in calcareous plates which only allow lateral movements. In some species these arms are covered in spines while in others they are smooth. Broken arms regenerate in a similar way to other armed echinoderm groups. The species so far confirmed in Anguilla are described below, but it is very possible that others exist. Their secluded nature makes sightings and also identification very tricky. Most species are more likely to be seen at night and are very sensitive to light.

Common Name – **Sponge Brittle Star**

Scientific Name – *Ophiothrix suensonii*

Description – The arms have a dark central line running along them and are covered by many thin, sharp-looking spines. Colours vary from yellow to grey to black. As the common name suggests, it is often found living in association with different species of sponge, particularly the Lavender Rope Sponge (*Niphates erecta* - page 34), Pink Vase Sponge (*Niphates digitalis* - page 32) and Branching Vase Sponge (*Callyspongia vaginalis* - page 31), although is also sometimes found on fire coral (page 40) and certain species of Gorgonians (page 65-68). Common, with a disc size up to c.1.5 cm and arms up to c.12 cm in length.

Common Name – **Blunt-Spined Brittle Star**

Scientific Name – *Ophiocoma echinata*

Description – The arms are relatively thick and covered with blunt-looking spines. Colours vary but usually dark, occasionally with a mottled pattern. Hides under rocks and in dark recesses, retreating rapidly if disturbed. Common, with a disc size up to c.3 cm and arms up to c.9 cm in length.

Common Name – **Ruby Brittle Star**

Scientific Name – *Ophioderma rubicundum*

Description – Predominantly red, although bands may be present on arms, and disc may be patterned with similar colours. Usually, this species is only observed at night when it can be locally common, for example in some rocky parts of Meads Bay. This species grows to a very large size, making them the largest brittle stars in Anguillian waters.

Notes on other brittle star species: Due to their cryptic nature a number of other species are likely to be present in Anguillian waters but have yet to be confirmed here. For example: The **Reticulated Brittle Star** (*Ophionereis reticulata*), the **Circle Marked Brittle Star** (*Ophioderma cinereum*), and the **Banded Arm Brittle Star** (*Ophioderma appressum*). Evidence for the existence of these species comes from remains found washed up on certain beaches (see page 215).

Order – Phrynophiurida: Basket Stars

Basket stars are splendid creatures whose five arms repeatedly subdivide into numerous smaller branches resembling plant roots. Unlike brittle stars, basket star arms lack calcareous plates, so can be moved in all directions. During the day they wrap themselves up tightly for protection and so are difficult to see, but at night they climb to the top of nearby structures and spread out their arms in a basket-like arrangement to filter plankton out of the water. They transfer the captured food to the central mouth using fine spines and their tube feet. In Anguilla only one species of basket star has been confirmed.

Common Name – **Giant Basket Star**

Scientific Name – *Astrophyton muricatum*

Description – Numerous thinly branched arms that by day are tightly curled up but at night are spread out in a filter-feeding basket arrangement. Colour varies from orange to dark brown. Occasional, with a small disc of c.4 cm in diameter, although arms can span up to c.50 cm.

Class – Echinoidea: Sea Urchins

Sea urchins are probably the largest group of echinoderms in the Caribbean, with some species living under the sand whilst others are quite conspicuous reef dwellers. They possess a skeleton of ten fused calcareous plates known as the test, which is usually covered with numerous articulated spines and the characteristic tube feet. The group can be split into three distinct sections based on their morphology, which consists of the classic domed urchins (belonging to approximately ten different Orders), the heart urchins and the sand dollars (including the sea biscuits).

Order – Various: Domed Urchins

The domed urchin is the largest group of sea urchins and contains those species with the classic spiky urchin morphology. There are at least ten different Orders belonging to this group, although only five species are commonly seen in Anguillian waters. They have spherical bodies with long protective spines and tube feet. The anus is on the top of their test and their mouth underneath. The mouth is known as an 'Aristotles Lantern', being a complex lantern-shaped arrangement of five interlocking teeth that the urchin uses to scrape algae off the reef. The spines of certain species are long, sharp and brittle, snapping off easily under the skin even if barely touched, while others are short, blunt and less threatening. Although puncture wounds are painful, no species in Anguilla are known to be toxic, and as the spines are made of calcium and protein any fragments left under the skin will slowly be dissolved by the body. However, wounds should still be treated to prevent infection, and to aid dissolution hot water can be applied to help denature the protein.

Interestingly, some species of sea urchin are considered keystone species, and viewed as essential parts of the reef ecosystem, being responsible for sculpting its very nature. One example of this is the Long-Spined Sea Urchin (*Diadema antillarum*) which can be found in large numbers on shallow reef environments and grazes algae to such an extent that it produces bare areas known as 'urchin barrens'. The grazing of rapidly growing algae in such a way gives slowly colonising hard corals (page 51-64) a chance to settle and grow which would otherwise have a much harder time establishing themselves. The monumental effect that this grazing has on the reef ecosystem was highlighted in the early 1980s when a virus spread through the Caribbean and devastated populations of *Diadema antillarum*. What followed was a rapid overgrowth of algae on reefs throughout the region. Damage caused by subsequent hurricanes and bleaching events could not be naturally repaired by recruiting corals because they were outcompeted by algae. This combined with an increased nutrient load from terrestrial runoff and pollution essentially feeding the algae has led to a 70% decline of coral cover in the Caribbean over the last 30 years. This topic will be discussed in more detail in Chapter 6 (page 220-230).

Common Name – **Long-Spined Sea Urchin**

Scientific Name – *Diadema antillarum*

Description – Has numerous long, thin, sharp spines. Is usually black although some individuals have a number of whitish spines due to lack of pigment. Juveniles also often have white and black banded stripes. Usually hides in recesses during the day, but where populations are dense aggregations can be commonly seen out in the open. Once abundant, this species was almost entirely wiped out by a virus that spread throughout the Caribbean in the 1980s. Although finally recovering, distributions are still patchy, but a few areas do have populations close to what it is thought they were before the virus struck. Body size may be up to c.10 cm with spines almost 20 cm in length. Do not touch!

Common Name – **Magnificent Urchin**

Scientific Name – *Astropyga magnifica*

Description – Has numerous long, thin, sharp spines which are usually striped and grouped together in clumps. Colour varies although is usually dark with large white segments bordered by bright blue spots. A characteristic fleshy 'bag', known as the dorsal sac, extends through the anal opening from the top of the urchin. Considered uncommon, but as the species prefers deep-water habitats actual abundances are not currently known. Body may grow over 12 cm in width with spines c.10 cm long.

Common Name – **Reef Urchin**

Scientific Name – *Echinometra viridis*

Description – Has relatively thick, tan-coloured spines with a greenish hue and lavender tips. Spines have a white ring around the base and are attached to a deep red body. Usually found in shallower regions where they hide in recesses during the day. Uncommon. The body is usually no larger than 5 cm with spines approximately 4 cm in length.

Photo Credit Stuart Wynne

Common Name – **Rock Boring Urchin**

Scientific Name – *Echinometra lucunter*

Description – Has short, relatively thick pointed spines that are usually black. Can be found in abundance on very shallow rocky environments where it bores into the limestone to make protective hideaways. Body size is usually no larger than 5 cm with spines c.3 cm in length. Although spines can produce a puncture wound they usually remain intact, leaving only a small amount of pigment under the skin. This species is a significant bioeroder and is responsible for much of the jagged limestone formations in Anguilla's coastal zone.

Photo Credit Stuart Wynne

Common Name – **Variegated Urchin**

Scientific Name – *Lytechinus variegatus*

Description – Has short, relatively thick pointed spines and obvious tube feet. Body and spines are usually green to white in the Caribbean, although can vary in colour substantially, from all white to deep purple. Uncommon: the pictured specimen was the first confirmed sighting in Anguilla (taken in Meads Bay). Usually no larger than 5 cm, with the ability to camouflage by covering themselves in algae and other debris.

Photo Credit Stuart Wynne

Common Name – **West Indian Sea Egg**

Scientific Name – *Tripneustes ventricosus*

Description – Is densely covered with short white spines and a dark reddish-black body. Often covers itself with debris for camouflage or protection from sunlight. Found on a variety of habitats, but seems to prefer seagrass areas. Although this species is harvested by fishers in many parts of the Caribbean, this practice is not commonly known to happen in Anguilla. This is probably because they are not particularly abundant. Occasional, but can be locally common. Body may be over 12 cm in diameter, with spines not usually exceeding 1.5 cm.

Common Name – **Slate-Pencil Urchin**

Scientific Name – *Eucidaris tribuloides*

Description – Has thick blunt spines that are often covered with a fine layer of algae and sediment. Body is usually reddish and has a pretty arrangement of white plates around the base of the spines. Occasional, but often overlooked because it wedges itself tightly into recesses during the day. The body is usually no larger than 5 cm with spines of approximately equal length.

Order – Spatangoida: Heart Urchins

Heart urchins are so called because of the resemblance they share with their namesake. They are irregular in shape, being more oval than the spherical domed urchins, with their mouth situated at the front of their flat underside. The anus has moved from the top of the dome to the back underside, features which all combine to produce a more 'aerodynamic' shape that aids their infaunal burrowing lifestyle. The mouth lacks an Aristotles Lantern as possessed by the previously described domed urchins. Although most species spend the majority of their life buried under the sand, they sometimes emerge at night. Their white tests are however commonly found by divers and snorkellers once the urchin has died, lost its spines and sifted naturally up to the surface of the sand. There is only one species currently confirmed in Anguillian waters.

Common Name – **Red Heart Urchin**

Scientific Name – *Meoma ventricosa*

Description – Deep red and dome-shaped with an indented five-petal flower design on its upper surface. Usually remains under the sand, but may emerge at night, otherwise only being observed on death when it naturally rises to the surface of the sand. Prefers sandy bottoms and seagrass areas where it is probably abundant, although this status is hard to quantify due to their infaunal nature. Body can reach up to 15 cm in diameter, with short spines no longer than 1 cm. Is quite similar in appearance to the Inflated Sea Biscuit (page 102), but this species does not burrow under the sand and has a raised petal design rather than an indented one.

Order – Clypeasteroida: Sand Dollars & Sea Biscuits

Sand dollars and sea biscuits are disc-shaped irregular echinoderms that have the classic five-petal design on their back similar to the heart urchins (previous). Unlike the heart urchins, but like the domed urchins, their mouth has an Aristotles Lantern which is centred on the underside and has the anus behind it. With this design, complemented by very short spines giving their surface a fuzzy appearance, they are very well adapted for a burrowing existence. Most species live exclusively under the sand and are rarely seen alive on the surface, only being uncovered after death when they become the classic sand dollar popularly collected by tourists.

Common Name – **Six-Keyhole Sand Dollar**

Scientific Name – *Leodia sexiesperforata*

Description – Extremely flat disc-shaped body with fine spines and six thin oval slots. Has the classic five-petal design on its upper surface, although this feature is less obvious than on the Five-Keyhole Sand Dollar (following). Usually remains under the sand, although may emerge at night, otherwise only being observed on death when it naturally rises to the surface of the sand. This species prefers sandy bottoms where it is probably common, although this status is hard to quantify due to their infaunal nature. Maximum size c.10 cm.

Notes on other sand dollar species: Living under the sand means that Sand Dollars and Sea Biscuits are rarely seen alive and, as such, the species in Anguillian waters are likely more diverse than presented here. Other members include: The **Inflated Sea Biscuit** (*Clypeaster rosaceus* - right), the **Sand Dollar** (*Clypeaster subdepressus*), and the **Five-Keyhole Sand Dollar** (*Mellita quinquiesperforata*). This latter species looks almost identical to *Leodia sexiesperforata* apart from a missing 'keyhole' near the leading edge. Evidence for the presence of these species in Anguilla comes from remains found washed up on certain beaches (see Chapter 5 - page 209-219).

The Inflated Sea Biscuit

Class – Holothuroidea: Sea Cucumbers

Sea cucumbers are sausage-shaped and initially seem to be the odd-one-out among the echinoderms. They have shell plates so reduced in size that they are no longer visible – instead being buried within their leathery skin. They also have no external spines or arms, although possessing a limited amount of tube-like feet (podia). The radial symmetry of the five body sections classic to echinoderms is also not visible in these animals, but is part of their internal structure and hence secures them a place within this Phylum. Some species can only be identified by dissection, with the internal plates often being key, although the two species listed are easily distinguishable. There are the currently five species confirmed to be present in Anguillian waters, but some are nocturnal and hard to observe.

Some species of sea cucumber are active during the day, and crawl across the surface scooping up organic matter through their mouth at the front and excreting waste through their anus at the rear. This excreta is characteristic of sea cucumbers and, where they are abundant, is a common feature on the sea bed (see top left corner of first image for Three Rowed Sea Cucumber on following page). Sea cucumbers probably have two of the most curious habits of all invertebrates. The first is that if threatened they excrete some of their internal organs through their mouth (some believe this is a strategy allowing escape while the predator feeds on their entrails), and the second is that they breathe through their anus.

Common Name – **Donkey Dung Sea Cucumber**

Scientific Name – *Holothuria mexicana*

Description – Dark brown to black with deep conspicuous creases that give it the resemblance of its namesake. The underside is pale with scattered short podia. Occasional, although may be locally common, preferring sandy areas and seagrass beds close to reefs. Maximum size c.35 cm.

Common Name – **Three-Rowed Sea Cucumber**

Scientific Name – *Isostichopus badionotus*

Description – Colourations vary greatly (compare two images to the right) but are usually a combination of dark brown or tan. Has knobbles on its back that are often a different colour from the rest of its body and three rows of podia on its underside. Uncommon to occasional, preferring a variety of habitats from sandy bottoms to rubble areas near reefs. Maximum size c.40 cm.

Notes on other sea cucumber species: The full extent of the sea cucumber community has yet to be established in Anguilla, although it is known that species other than those mentioned above are present. For example, the Furry Sea Cucumber (*Astichopus multifidus*) has been confirmed here (see image on right). Also, the **Beaded Sea Cucumber** (*Euapta lappa*) was recently sighted near the wreck of the M.V. Oosterdiep. Another confirmed member is a rather strange creature called the **Tigers Tail Sea Cucumber** (*Holothuria thomasi*) that lives in protective recesses within the reef. At night it becomes active and stretches out to sweep the surrounding area for food, while its posterior end remains firmly attached within the recess. If disturbed it rapidly retracts its body back into the reef, remaining there until the threat has passed.

The Furry Sea Cucumber

3.10 Phylum – Chordata: Chordates

The majority of chordates are back-boned animals (vertebrates) as described in Chapter 4 (page 106-208), although a small number of them, the urochordate subgroup, lack a backbone but share some other features. Most notable of these features is the presence of a tail at some point in their lifecycle (often during larval stages), a central nerve cord, and a notochord. The notochord is a support structure for the central nerve cord, an arrangement that is replaced by bones in the vertebrates. In Anguillian waters, this grey area between vertebrate and invertebrate classification is represented by the tunicates (below) which, based on their visible body plan, most definitely belong within the current chapter.

Class – Ascidiacea: Tunicates

Tunicates get their name from their cellulose body covering known as the tunic and are among the most common marine invertebrates. Despite this, they are often overlooked because of their small size and often cryptic nature. Most species live attached to a firm surface, filter feeding through two siphons. One of these siphons acts as an intake, letting in water that is then pumped through a gill net removing both oxygen and food. The water then leaves via the other siphon. These siphons can be opened and closed by muscles, a feature which helps to distinguish the larger species from sponges.

The Bulb Tunicate (*Clavelina* sp.)

Tunicates have a variety of growth forms. Most of the larger species are solitary, although many of the smaller ones grow in association with others and can be joined communally at the base. Some associations are even more intimate, where many individuals are embedded in a common tunic, or even share out-flowing siphons.

Common Name – **Giant Tunicate**

Scientific Name – *Polycarpa spongiabilis*

Description – Large soft but firm body with obvious siphons. Animal is usually covered with algae and lives in reef recesses although the inside of the siphons are orange which can make the creature stand out. Inside the in-current siphon is a ring of un-branched tentacles. Considered occasional, although its cryptic nature makes this status very hard to quantify. Maximum size c.10 cm.

Common Name – **Painted, Bulb & Bell Tunicates**

Scientific Name – *Clavelina* sp.

Description – Members of this Genus are characterised by a transparent to translucent tunic with a beautiful painted design outlining the siphons and other body features. Colourations vary. They grow in groups and are considered common but being transparent, small and able to grow inside reef recesses, this status is hard to quantify. Maximum size c.1.5 cm. The pictured specimens are probably a colour variety of the **Painted Tunicate** (*Clavelina picta*), although identification between species is extremely difficult.

Notes on other tunicate species: The species described above are only a selection of the species found in Anguillian waters, as their cryptic nature and difficulties with identification mean a detailed list is not possible to produce currently. Of those likely to be present, the most probable members of the tunicate community are the **Mangrove Tunicate** (*Ecteinascidia turbinata*), the **Encrusting Social Tunicate** (*Symplegma viride*), the **Flat Tunicate** (*Botrylloides nigrum*) and the other encrusting tunicates, most notably from the Genera *Distaplia*, *Diplosoma* and *Didemnum*.

The Mangrove Tunicate (*Ecteinascidia turbinata*)

The Encrusting Social Tunicate (*Symplegma viride*)

Class – Thaliacea: Pelagic Tunic

Pelagic tunicates are free-swimming tunicates found in open-water. They can be solitary or form chains and occasionally swim over reefs. The in-current and ex-current siphons are at opposite ends of the animal and muscle contractions push water through, thus producing locomotion. Being transparent, they are very hard to see and identify, not just to species level but also to Genus. For this reason they are only discussed here briefly and generically. The pictured specimen (below) remains unidentified.

The Class Thaliacea is the main group of pelagic tunicates within the Subphylum Urochordata, and contains the species most likely to be encountered in Anguilla (Order Salpida). Salps, as they are commonly referred to, have a complex lifecycle with alternating aggregate and solitary generations. Individuals within aggregates are hermaphrodites, and usually begin life as females. These are fertilized by older males from another chain, with the resulting embryos developing into the solitary phase, which reproduces asexually to form a subsequent aggregated generation. Salps play an important role in marine ecosystems as they can form massive clusters containing millions of individuals.

An unidentified Pelagic Tunicate, commonly known as Salps

Chapter 4 - Vertebrates

This chapter is dedicated to the vertebrate (back-boned) creatures. The vertebrate group consists of all fish, reptiles, birds and mammals. By far the largest group among the marine vertebrates are the fish, with over 185 species described in the following pages. Accompanying the fish in this group are Cetaceans (marine mammals page 204-206) and a small number of reptiles, represented by the sea turtles (page 206-208). Vertebrates all belong to the Phylum Chordata and form a Subphylum of their own known as Vertebrata. The chordates not only consist of vertebrates, but also more primitive creatures such as lampreys (fish-like organisms that are not featured in this book) and tunicates. Tunicates are placed within the Phylum Chordata because as larvae they possesses many standard chordate features (page 104-105), although being entirely soft-bodied as adults they look completely out of place within it and hence are part of the previous chapter (invertebrates). This means Phylum divisions are split across two chapters, which is a slight taxonomic conundrum, but one that is unavoidable thanks to the wonderful diversity of life found in the oceans.

4.1 Class – Pisces: Fish

Introduction

Although fish were formally grouped under the Class Pisces, as the above heading suggests, in recent years this categorisation has changed with Pisces being promoted to the level of Superclass, and modern fish are now generally thought of in terms of three separate Classes:

- Agnatha (jawless fish) - Not generally found in shallow water areas.

- Osteichthyes (bony fish) – Sometimes generically known as Teleosts, although this is a slightly smaller taxonomic group.

- Chondrichthyes (cartilaginous fish) – Sometimes generically known as Elasmobranchs. This is also a slightly smaller taxonomic group.

In Anguillian waters the dominating Class is that of the bony fish, which consists of virtually all the commonly occurring species seen swimming around the reefs. The cartilaginous fish are less abundant although equally important and comprise all sharks and rays. Cartilaginous fish differ from bony fish in a number of ways, most notably because their skeleton is made solely out of cartilage rather than typical bones. Other differences include the fact that sharks and rays do not posses scales, instead often having rough sandpaper-like skin. They also lack the swim bladder which gives bony fish control over their buoyancy, the result of which is that sharks have to maintain themselves in the water column via hydrodynamics in a similar way to an aeroplane flying, which explains their characteristic shape. If a shark stops swimming it quickly sinks to the bottom. Many rays have taken full advantage of this and live almost exclusively on the sea bed. The jawless fish, although probably present in deeper areas, have not yet been confirmed in Anguilla's shallower zones that are visited by divers and snorkellers.

The fish described in this section have been divided into Families, and assembled in slightly larger groups based on body plans and/or life history traits. This method of classification differs from the previous section as the majority of these Families all belong to the Order Perciformes (Perch-like fishes), which is complicated to split into smaller groups and thus not suitable for referencing. The morphological grouping presented here allows a fish spotter to quickly identify those individuals seen during a recent dive or snorkelling trip. It follows a general pattern beginning with fish possessing a classic body plan that are mostly silver in colour, and slowly expands into the somewhat more exotic:

Silver Swimmers – Including jacks, mackerels, barracuda and porgies

Sloping Head and Classic Body Plan – Grunts and snappers

Disc and Oval Bodies – Angelfish, butterflyfish, surgeonfish, damselfish & chromis

Large Lips and Heavy Body – Hamlets, groupers and other sea bass

Swims with Pectoral Fins – Parrotfish and wrasse

Large Eyes and Reddish Body – Including squirrelfish and cardinalfish

Small Bottom-Dwellers – Gobies, blennies, and jawfish

Odd-Shaped Bottom-Dwellers – Including scorpionfish and lizardfish

Odd-Shaped Swimmers – Including pufferfish, boxfish, goatfish and triggerfish

Eels – Morays, snake eels and conger eels

Sharks and Rays – Including carpet sharks and stingrays

General Terminology

The main body parts of a fish are depicted in the following schematic diagram, and are referred to throughout the section. Sometimes there are slight variations on these names where the ventral fins may be called the pelvic fins and the caudal fin discussed as a tail fin. In some cases the dorsal fin is split in to two or even three separate parts. Furthermore, if the dorsal fin is continuous the fore-dorsal part can be of a distinctly different structure to the rear-dorsal part. There may also be small finlets between the dorsal and caudal fins. Such characteristics can sometimes be key to identifying the differences between species.

Bony fish have sense organs that may not be familiar to many of us. One of these is the lateral line, which consists of a row of fluid-filled bags beneath the skin with a small, sensitive, hair-like structure inside. These detect pressure changes in the surrounding water thus acting as a warning device against predators, and helping with perception of the surrounding topography – a very useful sense in strong surges near sharp corals. Other sense organs include the cirri and barbels which are rarely present together on a single fish species. Goatfishes (page 194-195) for example possess barbels but not cirri, and many blennies (page 174-175) possess cirri but not barbels. Both structures are sense organs generally used to detect buried food (in the case of barbels) or floating food particles (in the case of cirri). Again, the presence or absence of such structures can be very useful when differentiating between species.

Another key feature to look out for when identifying fish is their colourations. Interestingly, colour itself is not a straight forward identification feature as many species of fish can change colours rapidly and dramatically, at times turning from a dark, almost black colour to a bright white colour win the blink of an eye. This is possible through the use of chromatophores - cells with many small pigment-filled sacks that can be expanded or contracted to make different colours dominant. Unfortunately for the casual observer the same thing goes for the patterns which the colours form, where a seemingly featureless individual can suddenly posses a set of dark stripes. These patterns and colour changes can be learnt and become key during identification, especially when combined with elements that remain unchanged. Such features are discussed throughout this section, with their broad definitions defined on the following page.

Body colouration pattern definitions used in this book:

- Stripes – Horizontal line marking

- Bands – Diagonal line markings

- Bars – Vertical line markings

- Spot – Well defined circular marking

- Ocellated Spot – Spot with a ring around it of another colour

- Blotch – Poorly defined irregular marking

- Speckles – Small spots similar to freckles

Each fish species listed in this section is also given a maximum size. This size has been taken from scientific literature and other sources and does not relate to sizes seen in Anguillian waters. These sizes, therefore, are the maximum the fish could reach were it to die of old age, an event which rarely happens these days in our world of commercial fishing and habitat destruction. It is however still a useful ID feature because if the fish in question is larger than the size listed in this book, it can be deleted from the list of suspects. As a rule of thumb, the size of an adult fish seen in our waters is usually a little less than half the maximum size quoted throughout this section, although obviously there are some exceptions.

Finally abundance, distribution and short notes on behaviour are given which again can be a useful way to decide if your identification skills are reaching the correct conclusions. For example, a fish that is said to be rare is very unlikely to be seen in large numbers or on a regular basis. Similarly, if a species is described as living on a reef, or only in shallow waters, it is unlikely to be seen on a deep sandy bottom. Finally, behaviour can also be a useful indicator, for example the similar Blue Chromis (page 143) and juvenile Creole Wrasse (page 162) can be identified easily, as the former generally remains swimming around the same area of habitat whereas the latter moves actively about the reef.

Silver Swimmers

Family – Istiophoridae (Sailfishes & Marlins) and Xiphiidae (Swordfishes)

Known collectively as Billfish due to their prominent bills (rostra), these large species are all exclusively pelagic and highly migratory. They are found worldwide, although many species favour tropical and subtropical waters. Their preferred habitat means they are unlikely to be seen while diving or snorkelling, although occasionally they may be found washed up on shore. For example, the pictured specimen (below) was found on Savannah Bay by Fisheries Officers in April 2007: based on its injuries this was probably due to a boat collision. Aside from this **Blue Marlin** (*Makaira nigricans*) other Billfish species found in Anguillian waters will include: White Marlin (*Tetrapturus albidus*); **Atlantic Sailfish** (*Istiophorus albicans*); and **Swordfish** (*Xiphias gladius*). As an interesting note, the Atlantic Sailfish is the fastest of all fish, reaching speeds of up to 110 km per hour. All species within this group are popular among sports fishers.

A Blue Marlin found washed up on Savannah Bay in 2007 after a presumed boat collision

Family – Coryphaenidae: Dolphinfishes

With only two recorded species in this Family, and only one of these known in local waters, the dolphinfishes comprise a small yet important group of fish for Anguilla. Known as Mahi-Mahi, this species is popular in the island's restaurants and thus targeted by local fishers. This large, fast swimming pelagic fish is also sought after by sports fishers who enjoy the challenge of landing a large bull. They are moderately slender and flattened sideways, with a slightly protruding lower jaw, massive blunt head, and a long dorsal fin without spines which extends from close behind the head to near the base of the caudal fin. They also have a long anal fin that extends from the mid-body to the base of the caudal fin.

Common Name – **Dolphinfish**

Scientific Name – *Coryphaena hippurus*

Local Name – Mahi-Mahi

Visual ID – Brilliant silver with long continuous dorsal fin from above eye to base of caudal fin. Males (bulls) have a very blunt head and display yellow-green-blue spots and blotches. Females have a rounded head with a more uniform blue iridescent sheen. Maximum size c.1.5 m.

Abundance, Distribution & Behaviour - A common pelagic species often targeted by fishers in Anguilla, although it is rarely seen over reefs. Most likely to be observed during a wall dive at Dog or Scrub Island. Swims rapidly in open-water, often in groups containing one or two bulls and numerous females.

Family – Carangidae: Jacks

Jacks are almost entirely silver and can reach moderate to large sizes. They usually have large eyes and are thin with small tail bases and a deeply forked tail. Due to this design they are generally strong, fast-swimming predators of the open sea although schools sometimes pass over reefs in search of food. A few species, especially younger individuals, can be seen regularly on the reef. Some of the offshore dive sites around Anguilla offer a great chance to see these impressive fish swimming in large schools, especially Anguillita, Dog Island, Scrub Island, and many of the wrecks. Other than the species listed following, others include the **Crevalle Jack** (*Caranx hippos*) and the **Yellow Jack** (*Caranx bartholomaei*), although these spend most of their time away from reef areas, preferring a more pelagic environment.

Common Name – **Rainbow Runner**

Scientific Name – *Elagatis bipinnulata*

Local Name – Rainbow Runner

Visual ID – Silvery yellow-blue with two blue stripes running length of body and a pale to brilliant yellow stripe in between. Head is more pointed and body more elongated than that of other jacks with small dorsal and ventral finlets present. Maximum size c.1 m.

Abundance, Distribution and Behaviour – Rarely sighted over reefs and considered more of a pelagic species. Can be sighted along drop-offs, near wrecks, or in other deeper areas, although has been seen making rapid passes in shallow sandy areas, for example offshore from Long Bay.

Common Name – **Bar Jack**

Scientific Name – *Caranx ruber*

Local Name – Cavalle

Visual ID – Silver with black border running along dorsal fin and down lower lobe of caudal fin. Can darken dramatically, becoming almost black. Maximum size c.65 cm.

Abundance, Distribution & Behaviour - Common on most reef habitats. Large schools are often seen at the offshore dive sites, especially wrecks, where they can mix with Blue Runners. Can be solitary, becoming an opportunistic feeder seen in association with goatfishes or shadowing stingrays while they rummage under the sand. Large individuals often stalk shoals of small silveries (page 123) which congregate in places such a Little Bay. Juveniles swim close to larger fish such a Barracuda in a similar fashion to pilotfishes, thus gaining protection. In this way they will often flirt harmlessly with snorkellers, swimming right in front of their mask with a persistence that means the only way to lose them is to get out of the water.

Common Name – **Blue Runner**

Scientific Name – *Caranx crysos*

Local Name – White Back

Visual ID – Bluish silver with dark tips on caudal fin. May have dark blotch near upper edge of gill cover and faint bluish bars on body. Breeding males can be a lot darker. Maximum size c.75 cm.

Abundance, Distribution & Behaviour - Occasional, but prefers more open water and so is rarely seen on reefs. Most commonly sighted in schools (sometimes with Bar Jacks) on offshore dive sites, wrecks, or in the water column above reefs or seagrass beds.

Common Name – **Horse-Eye Jack**

Scientific Name – *Caranx latus*

Local Name – Horse-Eye Cavalle

Visual ID – Silver with yellow caudal fin. Pectoral fin is clear, but scutes, tip of dorsal fin and upper caudal fin are usually dark. Large eyes. Maximum size c.75 cm.

Abundance, Distribution & Behaviour - Occasional, although may be locally common, especially on deeper reefs, the offshore dive sites and wrecks. They swim in open-water over reefs, sometimes forming very large schools. Schools may reside in one area for many years as can currently be seen at Anguillita. Juveniles are seen regularly in many shallow sandy areas including Captains Bay, Meads Bay, Rendezvous Bay, Cove Bay and Sandy Ground where they often swim amongst bathers.

Common Name – **Black Jack**

Scientific Name – *Caranx lugubris*

Local Name – Black Jack

Visual ID – Silver grey to almost black with steep forehead and long dark dorsal and anal fins. Caudal fin is also dark, as are mid-body scutes. Maximum size c.1 m.

Abundance, Distribution & Behaviour – Uncommon, although is often sighted in offshore waters, for example around Sombrero Island. Considered deep-water fish, and not a schooling species, usually travelling alone or in small groups/pairs.

Common Name – **Almaco Jack**

Scientific Name – *Seriola rivoliana*

Local Name – Almo Jack

Visual ID – Silver, with dark diagonal band running from front of dorsal fin, down across the eye to the lip. Fore-dorsal fin is usually long. Maximum size c.1 m.

Abundance, Distribution & Behaviour – Uncommon. A deep, open-water fish that rarely swims over reefs. Can be sighted at deeper offshore sites and wrecks. Usually solitary, although sometimes seen in schools.

Common Name – **African Pompano**

Scientific Name – *Alectis ciliaris*

Local Name – Moonshine

Visual ID – Silver with steep blunt forehead. Its scales are not obvious and often have bluish or greenish tints. Juvenile has long thread-like filaments trailing from dorsal and anal fins that become progressively shorter as the individual ages. Maximum size c.1 m.

Abundance, Distribution & Behaviour – Uncommon. An open-water fish that is rarely seen over reefs.

Common Name – **Permit**

Scientific Name –*Trachinotus falcatus*

Local Name – Ocean Cobbler

Visual ID – Silver with shades of blue-green on head. Dark caudal and pectoral fins, with tips of dorsal and anal fins dark also. Often has orange-yellow area on belly in front of anal fin. Maximum size c.1 m.

Abundance, Distribution & Behaviour – Uncommon to rare with only a handful of sightings over recent years. Usually solitary, swimming over sandy areas looking for molluscs on which they feed. On rare occasions may be sighted over reefs where they forage for urchins and crustaceans.

Common Name – **Palometa**

Scientific Name – *Trachinotus goodei*

Local Name – Cobbler

Visual ID - Silver with four dark body bars that often appear somewhat faded. Has long dorsal and anal fins that usually have dark tips. Maximum size c.45 cm.

Abundance, Distribution & Behaviour – Occasional although can be locally common. Small groups of younger individuals are seen regularly in shallow sandy areas including Captains Bay, Meads Bay, Rendezvous Bay, Cove Bay and Sandy Ground where they often swim amongst bathers. Schools of adults prefer more offshore regions although may be sighted in coastal areas close to open-water, an example being Katouche Bay. They are largely unconcerned by humans and will occasionally circle divers and snorkellers for extended periods.

Common Name – **Scad**

Scientific Name –*Decapterus* sp. & *Selar* sp.

Local Name – Jacks

Visual ID – The different species are difficult to distinguish underwater, although generally speaking they are bright reflective silver with a deeply forked tail and black spot at the top of their gill cover. Sizes vary depending on species with the **Round Scad** (*Decapterus punctatus*) reaching c.25 cm, **Mackerel Scad** (*Decapterus macarellus*) c.30 cm, and **Bigeye Scad** (*Selar crumenophthalmus*) sometimes attaining c.60 cm. Reportedly the most common species in Anguillian waters is the Bigeye Scad that can sometimes be caught in very large numbers in more offshore regions, for example around Dog Island.

Abundance, Distribution & Behaviour – Occasional depending on location. Can form large, rapidly swimming polarised schools. This behaviour is what makes species identification tricky. Although preferring open-water they can sometimes be seen racing over reefs, shallow bays, along walls and over wrecks. Are often seen along the sea rocks at Meads Bay.

Jacks and other important teleost fisheries: Of the smaller scale fisheries in Anguilla (thus excluding larger commercial fisheries such as long-lining for large pelagic species), jacks are up at the top of the list along with snappers (page 129-133), groupers (page 146-150) and mackerels (page 113-114). These four fish Families account for the majority of species directly targeted by commercial fishers in Anguilla, usually by using hook and line gear. Jacks and mackerels are predominantly caught by trolling, where a line is towed behind a moving fishing vessel, whereas groupers and snappers are demersal in nature and so caught with a stationary hook and line resting near the sea bed. To upscale catching snappers, some fishers employ what is known as a 'vertical long-line' where a weighted line with numerous hooks is dropped into deeper water and left for a few hours. Being demersal, snappers and groupers are also landed via fish traps that are used to target reef fish generically.

Family – Scombridae: Mackerels

Mackerels are silvery, and generally open ocean fish that only occasionally pass over reefs. They are strong, fast-swimming predators that are popular in the island's restaurants and thus targeted by local fishers. Due to their speed and strength these pelagic fishes are popular with sports fishers who enjoy the challenge of landing a large individual. They have two dorsal fins that fold into grooves reducing drag during bursts of speed. Behind these, and also the anal fin, there are a number of 'finlets' that are believed to keep the fish on course when the dorsal fins are tucked away. The base of the caudal fin is slender with two or more keels on each side that serve a similar purpose. This tail is widely forked, completing a design that facilitates extended periods of fast swimming. Tunas are also a member of this Family.

Photo Credit Steve Donahue

Photo Credit Paul Humann

Common Name – **Cero**

Scientific Name – *Scomberomorus regalis*

Local Name – Spanish Mackerel

Visual ID – Silvery, with a series of gold streaks accompanied by a number of small yellow-gold spots. Has an elongated body that can reach a maximum size of c.1.25 m.

Abundance, Distribution & Behaviour – Occasional on reefs. Considered an open-water fish which may be sighted in deeper offshore areas and on wrecks. Usually solitary, although sometimes seen in small groups/pairs. They are curious about divers, often allowing a close approach.

Common Name – **Wahoo**

Scientific Name – *Acanthocybium solandri*

Local Name – Wahoo

Visual ID – Silver, with sharply pointed nose and long cigar-shaped body. May display numerous dark bars when stressed. Caudal fin very widely forked. Maximum size c.2 m.

Abundance, Distribution & Behaviour – Rare on reefs, although may be sighted in deeper regions, walls or wrecks. They are sometimes seen in pairs but are usually a solitary open-water fish. Curious of divers often allowing a close approach.

Notes on other species from Family Scombridae: Of particular note is the **Little Tunny** (*Euthynnus alletteratus*), known locally as the Bonito when an adult, and Dumpling or Skip Jack when juvenile. They are silver, with diagonal wavy bands on their back and several spots below the pectoral fin similar to the classic Atlantic Mackerel. They reach a maximum size of c.1.25 m, and are primarily an open-water species, so uncommon on reefs. May be sighted occasionally near wrecks, or breaching the surface in shallow areas such as Sandy Ground when schooling. Another species of note is the **Spanish Mackerel** (*Scomberomorus maculatus*), which is similar in appearance to the Cero (above) but can be distinguished by yellowish spots rather than streaks. This species, thought to be rare in this part of the Caribbean, has been reported by fishers but no confirmed sightings made. It is suspected that these sightings are in fact juvenile King Mackerel (*Scomberomorus cavalla*), whose markings are very similar. The adult **King Mackerel** are often referred to as Kingfish which, although unlikely to be seen on the reef, are a species to note as a popular sports fish. Although the Little Tunny is known locally as Bonito, true **Bonito** (*Sarda sarda*), a favourite of local fishers, are also found in Anguillian waters but will rarely been seen while diving and snorkelling as they are primarily pelagic in nature. The same is true for Tuna species, which may be targeted by fishers but will seldom venture away from deeper offshore areas. According to published distribution maps, species that will be found in Anguilla's Exclusive Fisheries Zone that stretches 200 nm northwards from the mainland include: **Skipjack** (*Katsuwonus pelamis*); **Albacore** (*Thunnus alalunga*); **Blackfin** (*Thunnus atlanticus*); **Bigeye** (*Thunnus obesus*); and **Yellowfin** (*Thunnus albacares*). Commercially, Yellowfin Tuna is the most popular.

Family – Belonidae: Needlefishes

Highly reflective silvery fish with narrow slender bodies and long pointed jaws. They are considered common and found just below the waters surface above almost any habitat, although they prefer surf zones and shallow sandy bays. Usually seen in groups. The halfbeaks are visually similar to the needlefishes, although they belong in a different Family and so will be discussed separately. There are three species of needlefishes known to live in Anguillian waters, but identification is difficult and so all have been grouped here generically: The **Keeltail Needlefish** (*Platybelone argalus*) and **Houndfish** (*Tylosurus crocodilus*) are virtually indistinguishable, although the **Flat Needlefish** (*Ablennes hians*) can be recognised by faint dark body bars and a dark portion on the end of its dorsal fin.

Family – Exocoetidae: Flyingfishes & Halfbeaks

Although visually quite different the flyingfishes and halfbeaks are classified taxonomically in the same Family. Flyingfishes have elongated pectoral fins which they can extend when jumping out of the water enabling them to glide considerable distances. This behaviour is usually employed to avoid predators. Halfbeaks, rather similar to needlefishes, are differentiated by a short upper and extended bottom jaw. Again, as with the needlefishes, the flyingfishes and halfbeaks are difficult to distinguish from one another, so only the most common will be mentioned in any detail.

Common Name – **Atlantic Flyingfish**

Scientific Name – *Cheilopogon melanurus*

Local Name – Batfish

Visual ID – Whitish head and body with dark body bands on belly. Pectoral fins may have tints of blue, lilac, black and/or gold. Maximum size is thought to be c.25 cm (pictured is a juvenile).

Abundance, Distribution & Behaviour – Rarely seen on reefs although are common in open-water. During the day they stay relatively close to land-masses. Sometimes seen at night when they come over deep reefs to feed, and may be attracted to divers lights. Can often be seen breaching the surface and gliding great distances when startled by a passing boat.

Photo Credit Paul Humann

Common Name – **Ballyhoo**

Scientific Name – *Hemiramphus brasiliensis*

Local Name – Ballyhoo

Visual ID – Silver, with highly extended upper jaw. Upper lobe of caudal fin is yellow to orange which distinguishes it from the similar Balao (*Hemiramphus balao*) which has violet to blue colouration to both lobes. Maximum size c.40 cm.

Abundance, Distribution & Behaviour – Occasional, although can be common in shallow inshore-waters. Can be seen feeding in schools just below the surface over reefs, but more commonly in calm bays, lagoons and especially seagrass beds. This species is targeted for personal consumption by recreational fishers but reports of large individuals are now rare and catches much reduced. This suggests overfishing.

Family – Sphyraenidae: Barracudas

Barracudas are elongated silver fish with long jaws and numerous teeth. They have two low, widely separated dorsal fins and large forked tails. The larger, more solitary Great Barracuda is an easily identified member of the Anguillian fish community, and often targeted by fishers. It is reported to have the fastest predatory strike of any fish in the Caribbean. The two smaller barracuda species (following page) are more problematic to differentiate from each other, although attention to subtle colourations and surrounding habitat can help in this process.

Photo Credit Stuart Wynne

Common Name – **Great Barracuda**

Scientific Name – *Sphyraena barracuda*

Local Name – Barracuda

Visual ID – Silver with large jaw containing somewhat ominous teeth. Has a series of dark blotches on body and can darken to reveal side bands. Maximum size c.1.75 m.

Abundance, Distribution & Behaviour – Common. Drifts lazily over most reef habitats feeding on small fish. Opens and closes its mouth to assist respiration, a habit that can be rather intimidating. They are extremely curious and will closely approach divers, often following them around for extended periods. Unnerving as this habit is, there have been no reports of unprovoked attacks unless involving spear fishing or hand feeding. Usually solitary, although may be observed in small groups. Juveniles are seen in very sheltered areas of mangrove roots or seagrass beds.

Common Name – **Southern Sennet & Guaguanche**

Scientific Name – *Sphyraena picudilla* & *Sphyraena guachancho*

Local Name – Snook

Visual ID – The Southern Sennet is silver and slender with no obvious markings although they may have faint yellowish body stripes present. The Guaguanche on the other hand has a more obvious yellow or gold mid-body stripe. Maximum size: Southern Sennet c.45 cm; Guaguanche c.60 cm.

Abundance, Distribution & Behaviour – Uncommon, although the Southern Sennet is considered more common (pictured right). Southern Sennets prefer reefs or sandy areas within patch reefs, whereas the Guaguanche is more commonly sighted in shallow inshore-waters such as sand flats and seagrass beds. Although wary, both species can sometimes be closely approached and may briefly encircle a diver before slowly moving off. Big schools have been seen in the channel at Forest Bay and over sloping reefs in Shoal Bay East.

Family – Inermiidae: Bonnetmouths

Bonnetmouths are a small Family of fish that contains only two species, one of which has been confirmed in Anguilla. These slender fish feed mainly on plankton and so are usually associated with open-water, although they may form small mixed shoals over reefs with species of grunts (page 124-129). Bonnetmouths do not grow to substantial sizes and so are of little commercial interest.

Common Name – **Boga**

Scientific Name – *Inermia vittata*

Local Name - Boga

Visual ID – Silver with thin yellow stripes along elongate body. Has deeply forked caudal fin and fore-dorsal fin that connects with base of second dorsal fin. Yellow snout.

Abundance, Distribution & Behaviour – Rare. Occur most frequently on wrecks where species can sometimes be seen mixing with Tomtates (page 127). Pictured specimen was sighted on the wreck of the M.V. Catheley H.

Family – Albulidae: Bonefishes

Bonefishes are a group of fast-swimming silvery fish with a single dorsal fin, deeply forked tail, underslung mouth, and no obvious markings. They feed upon buried invertebrates, and can be extremely common in some parts of the Caribbean where they are the focus of a highly popular sports fishing industry (see note below). The attraction to fishermen is due to their eagerness to take bait and the fight they put up once hooked.

Common Name – **Bonefish**

Scientific Name – *Albula vulpes*

Local Name – Bonefish

Visual ID – Silver with short underslung mouth which does not extend beyond the eye. Maximum size c.1 m.

Abundance, Distribution & Behaviour – Rarely seen on reefs but may be sighted, when not feeding, on sand and coral rubble between shallow patch reefs. More commonly seen in the beach surf zone and over shallow sand flats on a rising tide. Have been sighted at Sandy Ground and Cove Bay, and are also reportedly common in Cove Pond and Long Pond.

Bonefishing: In some parts of the Caribbean bonefishing is a large industry, attracting sports fishers from around the globe. The Bonefish prefers to inhabit shallow sandy areas, and so territories with large stretches of this habitat take advantage of the resource to bring in tourist revenue. For example, the Bahamas and Florida have a thriving bonefishing sector, where the fish are caught using fly fishing gear and fishers stay in purpose-built bonefishing lodges. The fishers wade out into bays or utilise a shallow-draft boat, and subsequently release the catch (sometimes it may be retained and eaten). In Anguilla, there is no such industry as the necessary habitat is not extensive enough to support it, although bonefish may from time to time be caught and landed in shallow bay areas such as Sandy Ground or Cove Bay using regular hook and line fishing gear.

Family – Mugilidae: Mullets

Mullets are silver fish inhabiting inshore-waters, especially areas of sand, seagrasses and algae. Their foreheads are almost flat, with a curved lower jaw that gives them an almost upside-down appearance. They have small mouths and large scales with two widely separated dorsal fins. They may enter fresh water and often travel in huge schools. The most likely species to see in Anguilla is the **White Mullet** (*Mugil curema* - not pictured), which is silver with a dusky, almost red, tip of dorsal fin and margin of caudal fin. The pectoral fin also has slightly darkened base. It reaches a maximum size of c.35 cm, and is uncommon on reefs, preferring to inhabit areas of sand or seagrass and feed on organic material and small creatures found on plants and algae. They often swim around bays and inlets in large schools, with sightings made at Fish Hole Pond on Scrub Island, around the wooden jetty at Sandy Ground, and in Little Harbour.

Family – Sparidae: Porgies

Porgies are silver, with large steep heads giving them a high back profile. Their sometimes large eyes are set well above the mouth, and their bodies are often washed with tints of yellow and blue, although colourations can change quite dramatically. Porgies are usually solitary, although when one is sighted others are often seen in the vicinity. They stay near the bottom and, in Anguilla, are usually seen in deeper sandy areas, where they feed on shellfish and crabs. Identification between the different species is sometimes difficult, a fact that is not helped by the porgies habit of turning their backs on their observers.

Common Name – **Saucereye Porgy**

Scientific Name – *Calamus calamus*

Local Name – Porgy

Visual ID – Silver with yellow and blue washes, often this yellow colour is pronounced on the head and fore-body. Its common name derives from an obvious short, bluish, saucer-shaped line below the eye, but other porgies have a similar marking which makes identification tricky. Other subtle ID features include a small bluish blotch at upper base of pectoral fin and yellow corners to mouth. It can change colour rapidly and show a striped/blotched pattern. Maximum size c.40 cm.

Abundance, Distribution & Behaviour – Occasional on reefs close to sandy areas, often hovering motionless. Seems to favour deeper regions and is regularly seen on wrecks, with frequent sightings at Anguillita.

Common Name – **Jolthead Porgy**

Scientific Name – *Calamus bajonado*

Local Name – Porgy

Visual ID – Silver, with large eyes, long sloping snout and large mouth with thick lips. May have faint blue iridescence and can change colour dramatically to show striped or blotched pattern. A blue line directly under its eye is often visible. Maximum size c.60 cm.

Abundance, Distribution & Behaviour – Occasional on reefs close to sandy areas, often hovering motionless. Seems to favour deeper regions and is regularly seen on wrecks, with sightings made frequently at Anguillita.

Common Name – **Pluma**

Scientific Name – *Calamus pennatula*

Local Name – Porgy

Visual ID – Silver with yellow washes over its head, especially over its fore-back. Has short, blue, rectangular stripe behind the eye and bluish irregular stripes below. A semicircular marking, also bluish, is often present directly below eye. The Pluma can change colour dramatically to show a striped or blotched pattern. Maximum size c.40 cm.

Abundance, Distribution & Behaviour - Occasional on reefs close to sandy areas, often hovering motionless. Also favours seagrass beds. Often seen in deeper regions and on wrecks, with sightings made frequently at Anguillita and Limestone Bay.

Family – Gerreidae: Mojarras

Mojarras, especially when younger, are bright, reflective silvery fish. They have obvious scales and deeply forked tails. As juveniles they are rarely seen on reefs, instead inhabiting shallow sandy areas, especially within the surf zone. Older individuals begin to move to more open areas of sand, rubble and seagrass where they begin feeding on small invertebrates. Other than the two species listed following, the **Mottled Mojarra** (*Eucinostomus lefroyi*) has also been confirmed in Anguillian waters, seen swimming in the surf zone at Maundays Bay, and also seen landed in a fishers cast net at Island Harbour. It is bright silver in colour with a dusky area on its upper iris and further dusky areas on fore-dorsal and caudal fin margins, sometimes with faint body bars or mottling on its back.

Common Name – **Yellowfin Mojarra**

Scientific Name – *Gerres cinereus*

Local Name – Heado

Visual ID – Bright silver with yellow ventral fins. Pectoral fins may also be yellow. Indistinct vertical bars on body become more prominent with age and are a useful ID feature. Maximum size c.40 cm.

Abundance, Distribution & Behaviour – Occasional, although may be locally common especially in deeper reef regions close to sandy areas where they dig for small invertebrates. They are usually solitary but may be seen in small groups, for example at Anguillita, where a reasonably large population is present.

Common Name – **Flagfin Mojarra**

Scientific Name – *Eucinostomus melanopterus*

Local Name – Silver

Visual ID – Bright silver with black tip to dorsal fin and white stripe below. May have pale green wash over body. Maximum size c.15 cm.

Abundance, Distribution & Behaviour – Uncommon to occasional, with young individuals sometimes sighted in shallow sandy areas, especially in the surf zone of beaches such as Maundays Bay. They dart from spot to spot, abruptly stopping to rapidly feed, raising their dorsal fin from time to time.

Family – Elopidae: Tarpons

Tarpons are large-bodied silver fish that inhabit coastal waters, bays, estuaries and mangrove-lined lagoons, often entering fresh water. Interestingly, their swim bladder can be filled directly with air and permits the fish to live in oxygen-poor waters. In shallow waters their dorsal fins regularly breach the waters surface which, combined with their large size, gives them a somewhat shark-like presence. Tarpon are a highly appreciated sport fish famous for their spectacular leaps when hooked. The flesh is also highly prized despite being bony, and they are often targeted by fishers in Anguilla. Only one species is present in local waters

Common Name – **Tarpon**

Scientific Name – *Megalops atlanticus*

Local Name – Bass

Visual ID – Large silver fish with sizeable 'stainless steel' scales. Has an upturned mouth and small hump just behind the forehead. Deeply forked caudal fin. Maximum size c.2.25 m.

Abundance, Distribution & Behaviour – Occasional. Drifts around lazily by day in small to large groups, generally over reefs close to open-water where they feed actively at night on small fish. Often inhabit a specific area for years where they can be reliably observed, for example Anguillita, Limestone Bay and Shoal Bay East. Sometimes also sighted swimming under the wooden jetty at Sandy Ground. Their dried scales are sometimes used as guitar picks.

Family – Kyphosidae: Chubs

The Sea Chubs found in Anguillian waters are shaped rather like an American football. They are carnivorous, usually feeding on benthic invertebrates and swimming about the reef in small to large groups. It is not possible to differentiate the two species found in the Caribbean without counting their fin rays and gill rakers, hence they are discussed together below. Currently, it is not known whether local populations include one or both species.

Common Name – **Chub (Bermuda/Yellow)**

Scientific Name – *Kyphosus sectatrix/incisor*

Local Name – Chub

Visual ID –Silver to grey with oval body and dark caudal fin. Thin yellow to bronze lines may also be visible. Occasionally displays white blotches. Maximum size c.75 cm.

Abundance, Distribution & Behaviour – Occasional, although can be locally common or even abundant. Small to medium-sized schools swim about the reef, seeming to prefer locations close to deeper water. Dog Island and Scrub Island are particularly good places to find them in abundance.

Family – Ephippidae: Spadefishes

The Spadefish Family has only one member in the Caribbean and is a deep-bodied, compressed disc-shaped fish with a blunt snout often leading it to be misidentified as an angelfish. It visually differs, however, by having two dorsal fins and no gill spine. Although usually considered a pelagic species, in some parts of its range it resides in shallow marine and brackish waters of mangroves and harbours, frequently schooling in large numbers.

Common Name – **Atlantic Spadefish**

Scientific Name – *Chaetodipterus faber*

Local Name – Cobbler

Visual ID – Silver with dark bars that can pale dramatically. Body is oval-shaped somewhat resembling that of an Angelfish. Maximum size c.90 cm.

Abundance, Distribution & Behaviour – Uncommon to occasional, although is a regular visitor to many of the wrecks. Usually seen over reefs close to deeper regions or in open-water. Sometimes solitary but more normally in groups, this species is often observable at Anguillita.

Family: Small Silveries – Clupeidae (Herrings), Atherinidae (Silversides), Engraulididae (Anchovies)

These Families are almost indistinguishable from each other, especially to the average observer, not only because they are morphologically so similar, but also because they are usually small and move about rapidly in polarised, often mixed, schools. They have forked tails and are often transparent allowing fleeting glimpses of their internal organs and skeletons. These schools are often targeted by fishers who use them for bait, or fry them whole for food, hence their generic names: Baitfish and/or fries. Fries are caught from jetties and sea rocks by using a cast net. This round net is thrown skilfully onto the surface of the water above the school and its weighted edge sinks, encircling part of the school below. The slightly larger Redear Herring, locally known as Sprat, is caught in a similar way. It is discussed below separately because it is easily identifiable underwater.

Common Name – **Redear Herring**

Scientific Name – *Harengula humerali*

Local Name – Sprat

Visual ID – Small silver fish with faint lines and red to yellow spot on the upper end of gill cover. Snout and lower jaw may also be yellow. Maximum size c.20 cm.

Abundance, Distribution & Behaviour – Locally common, forming polarised schools in shallow bays and coastal waters. Common in nutrient-rich waters, especially near sewage outflows. Frequently seen and netted by fishers off the jetty at Island Harbour and have also been seen in huge schools under the cliffs near Little Bay.

Common Name – **Silversides** (Generic)

Scientific Name – Various

Local Name – Fries, Baitfish

Visual ID – Tiny silver fish with forked tails. May be less than 2 cm in length but can reach up to 15 cm depending on species.

Abundance, Distribution & Behaviour – Locally abundant, especially near under-cut cliffs, in caves and other sheltered reef areas. Swim in large polarised shoals, making for an extremely beautiful experience if seen while snorkelling. Little Bay is probably one of the most reliable places to view these shoals which usually consist of juvenile silverside, anchovy or herring species.

Sloping Head and Classic Body Plan

Family – Haemulidae: Grunts

Grunts are generally silver although most species have differing patterns of yellow, and sometime blue, stripes combined with other black colourations. They are closely related to the snappers but are generally smaller with less robust bodies and more deeply forked tails. Body shape does vary from species to species and, combined with stripe patterns, is useful for identification. Their name comes from the grunting noise they sometimes make, a sound that is produced when their swim bladder amplifies the sound made by grinding their back teeth. During the day some species form small to large schools that drift lazily over the reef and under overhangs. They feed nocturnally by scavenging sand flats and nearby seagrass beds for crustaceans. Juvenile grunts look very similar to each other making identification difficult, often forming large aggregations around coral heads of mixed species; for this reason they are discussed here generically. It is suggested that those interested in differentiating the juvenile grunts consult Humann & Deloach (1989).

Common Name – **Juvenile Grunts**

Scientific Name – *Haemulon* sp.

Local Name – None

Visual ID – Silvery with a series of light to dark stripes along their body and a dark blotch just before the often transparent caudal fin. Identification is made by comparing line patterns, but their small size and flitting movements often make this difficult. As individuals age the adult markings slowly start to develop but, depending on species, may not be obvious until c.10 cm in length. The White Margate, although from the Genus *Haemulon* does not follow this colouration pattern and lacks stripes.

Abundance, Distribution & Behaviour – Occasional to abundant depending on location and season. Usually form large shoals of varying sizes and species, normally positioned around a coral head or other protruding formation. Also found on seagrass beds. Juveniles can be recognised as grunts from as small as <1 cm.

Common Name – **French Grunt**

Scientific Name – *Haemulon flavolineatum*

Local Name – French Grunt

Visual ID – Yellow stripes on white to bluish-silver background. Stripes are horizontal over the lateral line and diagonal below it. All fins are yellow. Sometimes body can display two or more thick dark stripes. Maximum size c.30 cm.

Abundance, Distribution & Behaviour – Common on most reef habitats and can be locally abundant, forming large schools often in the shade of rocky formations (bottom image). Drift around lazily, although schools will swim away if approached. This is the most common grunt in Anguillian waters.

Common Name – **Bluestriped Grunt**

Scientific Name – *Haemulon sciurus*

Local Name – Grunt

Visual ID - Blue stripes over yellow to gold background with dark caudal and rear dorsal fin, which may have lighter margins. Maximum size c.45 cm.

Abundance, Distribution & Behaviour – Occasional to uncommon, although have been seen on most reef habitats, seeming to favour more protected areas. Usually solitary in Anguillian waters and generally nervous, moving into cover when approached.

Photo Credit Florent Charpin

Common Name – **Smallmouth Grunt**

Scientific Name – *Haemulon chrysargyreum*

Local Name – Grunt

Visual ID – Bluish silver, elongated body with five to six yellow stripes and yellow fins. Has noticeably smaller mouth than other grunts. Maximum size c.25 cm.

Abundance, Distribution & Behaviour – Uncommon, although can be common in certain areas, for example Sile Bay, Blowing Rock and west of Barnes Bay. Seem to prefer shallow reefs where they may be seen drifting in small schools usually in the shelter of protective formations.

Photo Credit Florent Charpin

Common Name – **White Grunt**

Scientific Name – *Haemulon plumierii*

Local Name – Grunt

Visual ID – Silver to white in colour with yellow and bluish-silver stripes on head. Large scales on body form a somewhat chequered pattern. Maximum size c.45 cm.

Abundance, Distribution & Behaviour – Uncommon, and usually seen solitary in Anguillian waters. Generally prefers patch reefs, and have been sighted in Forest Bay.

Photo Credit Stuart Wynne

Common Name – **Caesar Grunt**

Scientific Name – *Haemulon carbonarium*

Local Name – Grunt

Visual ID – Silver-blue with thin yellow stripes that can darken into thicker, deep bronze stripes. Dusky anal, caudal and rear dorsal fin. Maximum size c.35 cm.

Abundance, Distribution & Behaviour – Occasional although can be locally common. Probably, after the French Grunt (*Haemulon flavolineatum*), this is the most easily observed grunt in Anguillian waters (aside from schooling Tomtates (*Haemulon aurolineatum* - page 127) on many of the wrecks). Found on most reef habitats, and like other grunts may drift in small schools near rock formations and in recesses.

Common Name – **Tomtate**

Scientific Name – *Haemulon aurolineatum*

Local Name – Whipster

Visual ID – Silver-white with prominent yellow stripe running from snout, through the eye to the caudal fin. The upper body has another narrower stripe, interspersed with a varying number of pale thinner stripes. Often has a dark blotch on the caudal peduncle, although this is not an accurate ID feature as it can be absent. Maximum size c.25 cm.

Abundance, Distribution & Behaviour – Occasional, although can be locally common or abundant, forming large free-swimming schools over certain reefs and on wrecks. Seems to favour shallow reefs that slope into the deep, for example parts of Shoal Bay east.

Photo Credit Stuart Wynne

Common Name – **Cottonwick**

Scientific Name – *Haemulon melanurum*

Local Name – Cottonwick

Visual ID – Silver-white body with black stripe on dorsal fin running down and over the caudal fin. Has a dark stripe that runs from the snout, sometimes fading to yellow as it passes across eye and down towards caudal fin. Has a varying number of paler, thinner stripes over body. Maximum size c.30 cm.

Abundance, Distribution & Behaviour – Uncommon. Seem to prefer clear water reefs and are usually solitary, although may be seen in small groups.

Photo Credit Paul Humann

Common Name – **Striped Grunt**

Scientific Name – *Haemulon striatum*

Local Name – Grunt

Visual ID – Silver-white elongated body with approximately five yellow to brown stripes on upper body. The lack of stripes on its belly differentiates it from the Smallmouth Grunt which, considering its name, is rather confusing. Maximum size c.20 cm.

Abundance, Distribution & Behaviour – Uncommon, although are consistently seen in certain locations, for example north of Seal Island.

Photo Credit Paul Humann

Photo Credit Florent Charpin

Common Name – **Spanish Grunt**

Scientific Name – *Haemulon macrostomum*

Local Name – Grunt

Visual ID – Silver-grey with somewhat bold black stripes on upper body and a yellow-green dorsal hump. Also has a yellow saddle on the caudal peduncle, yellow pectoral fins and yellow borders to its caudal, anal and rear dorsal fins. Maximum size c.45 cm.

Abundance, Distribution & Behaviour – Occasional. Drifts alone or in small groups near secluded areas of reef. Prefers clear water although a number have been seen near Isaacs Cliffs where conditions are often quite turbid. Sightings have also been confirmed in and around the rocky reef areas close to Katouche Bay.

Photo Credit Florent Charpin

Common Name – **Sailors Choice**

Scientific Name – *Haemulon parra*

Local Name – Grunt

Visual ID – Silver-grey with black spots on scales forming vague body stripes. Fins are dusky and eye is relatively large with a golden hue around the pupil. Maximum size c.25 cm.

Abundance, Distribution & Behaviour – Uncommon to rare. Inhabits open reefs and, although it reportedly drifts in small schools elsewhere in the Caribbean, is usually solitary in Anguillian waters. This species is very difficult to approach, swimming rapidly away from any disturbance.

Photo Credit Steve Donahue

Common Name – **Porkfish**

Scientific Name – *Anisotremus virginicus*

Local Name – Porkfish, Porgy

Visual ID – Yellow-gold with silver body stripes. Probably the most strikingly coloured grunt. High back profile with a black band through eye and black bar behind gill cover. Maximum size c.35 cm.

Abundance, Distribution & Behaviour – Uncommon to rare. Usually solitary in Anguilla, although in other regions may school in huge numbers. Sightings have been made at Anguillita and on the offshore reef close to Crocus Bay.

Common Name – **Black Margate**

Scientific Name – *Anisotremus surinamensis*

Local Name – Thick-Lip Margate

Visual ID – Silver-grey with high back profile and large dark patch behind pectoral fin. Has dusky fins and dark central patches on many of its scales. Heavy lips. Maximum size c.60 cm.

Abundance, Distribution & Behaviour – Occasional, although can be locally common in areas such as Scrub Island and on many of the wrecks. Hides within reef recesses or under overhangs.

Common Name – **White Margate**

Scientific Name – *Haemulon album*

Local Name – Margate

Visual ID – Grey with high back profile and small pin-hole eye. Dorsal and caudal fins are usually dusky, the latter of which often has a bluish margin. This is the largest of the grunts with a maximum size of over 65 cm.

Abundance, Distribution & Behaviour – Occasional, although is commonly seen on many of the wrecks. May inhabit sandy areas between patch reefs and seems to prefer clear water.

Family – Lutjanidae: Snappers

Snappers are nocturnal predators that feed on crustaceans and small fish. Their common name derives from a habit of snapping their jaws when caught on a fishing line. The body shape is often somewhat oblong with a triangular head, and is commonly silvery-grey in colour. Similar colourations between species can make identification difficult at times, although subtle patterns usually make identification possible.

Once or twice a year, synchronised by the lunar cycle, mature snappers of certain species migrate long distances to reproduce at traditional spawning grounds. Others spawn in small resident spawning aggregations within their local area. Snappers share this large-scale aggregation behaviour with groupers, where migrations of more than a hundred miles occur over a period of days. Gradually, as fishers learnt about the locations and specific seasonality of such events, they could, in a very short space of time, remove the majority of mature snappers from whole areas. This has led to some snapper species being far less common today than recorded in the past.

The snappers described on the following pages are those most commonly seen while diving and snorkelling. Other deep-water species that live below safe-diving limits may be seen as landed catch on jetties around the island after being targeted by local fishers. These species include the **Red Snapper** (*Lutjanus campechanus*), the **Blackfin Snapper** (*Lutjanus buccanella*), the **Queen Snapper** (*Etelis oculatus*) and the **Vermilion Snapper** (*Rhomboplites aurorubens*). Juveniles of the two former species may occasionally wander close to shallow reefs, or can be mixed within shoals of other fish, so should not be discounted completely while fish-spotting. The young Red Snapper is pinkish-red, fading into silvery white on the lower body, with a dusky spot below the rear dorsal fin. The juvenile Blackfin Snapper is bluish-white to pale brown, with a dark blotch at the base of its pectoral fin, and brilliant yellow patches both below the rear dorsal fin and over most of its caudal fin.

Photo Credit Florent Charpin

Common Name – **Mutton Snapper**

Scientific Name – *Lutjanus analis*

Local Name – Buntail

Visual ID – Silver to grey with black spot on mid-body in line with the beginning of anal fin. May have faint body bars, and can darken dramatically. Can change to reddish brown, when blue lines under eyes become more obvious. Maximum size c.75 cm.

Abundance, Distribution & Behaviour – Uncommon. Drifts near bottom, usually over sand patches close to reefs, or on the reef edge. Most likely to be sighted on deeper reefs or close to wrecks.

Photo Credit Florent Charpin

Common Name – **Grey Snapper**

Scientific Name – *Lutjanus griseus*

Local Name – Bream

Visual ID – Silver-grey to reddish brown with no distinguishing features. Identification is usually by a process of elimination, although a dark band across the eye is often visible, especially in younger individuals (bottom image). Maximum size c.60 cm.

Abundance, Distribution & Behaviour – Occasional, but can be locally common on reef slopes close to deep water, for example Shoal Bay East. Younger individuals can be seen in a variety of habitats, especially in shallow inshore areas, rocky outcroppings and under jetties. Often seen in schools or may mix with other grunts and snappers.

Photo Credit Florent Charpin

Common Name – **Dog Snapper**

Scientific Name – *Lutjanus jocu*

Local Name – Dogteeth Snapper

Visual ID – Pale grey to reddish brown, with the ability to change colour dramatically. Has a pale triangular area under each eye. Its common name derives from two fang-like teeth which may be visible on the upper jaw. Maximum size c.1 m.

Abundance, Distribution & Behaviour – Uncommon, although is sometimes seen around wrecks and has been reported at Anguillita. Prefers shady areas.

Common Name – **Mahogany Snapper**

Scientific Name – *Lutjanus mahogoni*

Local Name – Mahogany Snapper

Visual ID – Silver to white with large eye and reddish margin on dorsal, anal and caudal fins. May have dark spot in line with anal fin, especially when younger. These features make it quite easy to confuse with the Lane Snapper (next). The dark spot may be ocellated in very young individuals. Maximum size c.40 cm.

Abundance, Distribution & Behaviour – Occasional, although can be locally common. Drifts alone or in small groups over reefs, seeming to favour deeper areas, with sightings at Anguillita and around many of the wrecks. Juveniles are often seen in shallow inshore areas.

Common Name – **Lane Snapper**

Scientific Name – *Lutjanus synagris*

Local Name – Sand/Grass/Virgin Snapper

Visual ID – Silver body. Pectoral, ventral and anal fins occasionally yellow. Has several faint yellow to pink body stripes and often displays a dark spot on mid-body. This spot, combined with a faint red margin to the caudal fin, makes it easy to confuse with the Mahogany Snapper (previous). Maximum size c.40 cm.

Abundance, Distribution & Behaviour – Occasional, although may be locally common forming shoals mixed with other grunts and snappers. Juveniles are often seen around the rocks in Little Bay area. Inhabits shallow to moderate depths on either reefs or seagrass beds but may be seen in deeper waters and on wrecks.

Common Name – **Schoolmaster**

Scientific Name – *Lutjanus apodus*

Local Name – Snaggertooth

Visual ID – Silver to copper colour with bright yellow fins and faint body bars (top image). These body bars and copper colourations are far more pronounced in juveniles (bottom image). Maximum size c.60 cm.

Abundance, Distribution & Behaviour – Occasional in Anguilla, probably due to fishing pressure. Historically drifted around in medium-sized schools but now are usually seen alone or with only one or two others. Juveniles are often seen inhabiting sheltered inshore rocky areas, shallow bays, and mangroves.

Common Name – **Yellowtail Snapper**

Scientific Name – *Ocyurus chrysurus*

Local Name – Snare

Visual ID – Silver with torpedo-shaped body and bright yellow stripe and caudal fin. May have yellow spots on upper body and yellow dorsal fin. Has a deeply forked tail. Can be confused with drifting schools of Yellow Goatfish (page 195). Congregational behaviour easily distinguishes the two species as Yellowtail Snappers swim more actively around the reef. Sometimes, however, these species mix so careful observation is required to identify correctly. Maximum size c.75 cm.

Abundance, Distribution & Behaviour – Common on most reef habitats but locally abundant in certain areas when schooling. Larger individuals are seen in groups over reefs (top image), more often those sloping into deeper waters or around wrecks. Juveniles are regular inhabitants of seagrass beds, and look like miniature versions of the adult. Medium-sized individuals swim alone or in loose aggregations over all reef habitats (bottom image).

Disc and Oval Bodies

Family – Pomacanthidae: Angelfishes

Angelfishes are one of the most decorative reef fish Families in the Caribbean. Their disc-shaped body, large size and graceful nature mean they are a real treat to spot. Their paucity in Anguillian waters is probably a result of overfishing. These days they are more commonly sighted in deeper water, so diving is the best means of attempting to find one. Adults have rounded foreheads with long dorsal and anal fins. Although some of the smaller species (not featured in this book) look quite similar to damselfish (later in this section), they can be morphologically differentiated by the presence of a spine protruding from the rear of their cheek across the lower gill cover. Sponges are their preferred food. Individuals from the Genus *Holacanthus* are protogynous hermaphrodites that often live in harems, where the largest harem female changes sex when the dominant male is no longer present. Angelfish from the Genus *Pomacanthus* do not change sex and individuals form extended, possibly lifelong, monogamous relationships with similarly-sized members of the opposite sex.

Common Name – **Rock Beauty**

Scientific Name – *Holcanthus tricolor*

Local Name – Swede

Visual ID – Black, with yellow fore-body, caudal fin and border to elongated dorsal and anal fins (top image). Lips are occasionally blue. The juvenile (bottom image), quite easily mistaken for a young Threespot Damselfish (page 140), is yellow with black ocellated spot on rear of body, which gradually expands with age. Maximum size c.30 cm.

Abundance, Distribution & Behaviour - Occasional, although can be locally common making this arguably the most easily spotted Angelfish in Anguillian waters. Somewhat shy, but they patrol established territories on the reef and chase off intruders quite aggressively. Juveniles are often seen hiding in reef recesses.

Photo Credit Stuart Wynne

Photo Credit Stuart Wynne

Common Name – **Queen Angelfish**

Scientific Name – *Holacanthus ciliaris*

Local Name – Angelfish

Visual ID – Brilliantly coloured with dark blue 'crown' on forehead and a yellow caudal fin. Body is blue overall but scales have varying degrees of yellow edging that provide some variety. Has yellow colouration on face, pectoral fins and ventral fins. The juvenile has similar brilliant colourations but with body bars and is very similar to the juvenile Blue Angelfish which has yet to be confirmed in Anguillian waters. Consequently the Townsend Angelfish – a hybrid of the two – has also yet to be confirmed. Maximum size c.45 cm.

Abundance, Distribution & Behaviour - Occasional. Can be found on most of Anguilla's reef habitats but seem to favour reef areas deeper than 5 m and are more common on wrecks. They swim about slowly and are somewhat shy, seldom allowing close observation.

Photo Credit Stuart Wynne

Photo Credit Stuart Wynne

Photo Credit Stuart Wynne

Common Name – **French Angelfish**

Scientific Name – *Pomacanthus paru*

Local Name – Angelfish

Visual ID – Black with yellow flecks on body scales. Has a yellow ring around eye and a yellow base to pectoral fin. The juvenile has three body bands, with a further band down snout and a yellow ring around caudal fin. This latter marking distinguishes it from the similar juvenile Grey Angelfish (next). As maturity is reached, these body bars fade but are sometimes still visible when quite large. Maximum size c.45 cm.

Abundance, Distribution & Behaviour - Occasional. Can be found on most of Anguilla's reef habitats, but seem to favour sloping reef areas deeper than 5 m and wrecks. They swim about the reef taking occasional bites from sponges. Adults are often in pairs and juveniles somewhat reclusive.

Common Name – **Grey Angelfish**

Scientific Name – *Pomacanthus arcuatus*

Local Name – Angelfish

Visual ID – Uniformly grey with pale yellow outer edge to pectoral fin. The juvenile, which is very similar to that of the French Angelfish (previous), has only a yellow band on its caudal fin, the outer margin being transparent. Intermediate stages often display a single faint white body band. Maximum size c.60 cm.

Abundance, Distribution & Behaviour - Uncommon in Anguilla although, potentially, can be seen on most reef habitats, especially those sloping into the deep. They swim about the reef taking occasional bites from sponges. Adults are often in pairs and juveniles somewhat reclusive.

Family – Chaetodontidae: Butterflyfishes

Butterflyfishes are small, often colourful fish with thin, round bodies rarely exceeding 10 to 15 cm in length. This makes them easy to distinguish from the similarly-shaped angelfish. Like some angelfish they are commonly seen swimming about the reef in pairs forming, what are believed to be, lifelong monogamous bonds. Their eyes are usually concealed by a dark bar which, combined sometimes with an ocellated spot on their rear quarters, is thought to confuse potential predators. Their spiny dorsal fin also serves as added protection. Using their protruding mouths, butterflyfish feed mainly on small worms, coral polyps and zoanthids, although each species tends to have its particular favourite. They are considered to be indicator species, where an abundance of individuals is believed to be a sign of a healthy coral reef habitat. Although adults of different species are easy to distinguish juveniles are sometimes similar which makes their identification rather more of a challenge.

Common Name – **Banded Butterflyfish**

Scientific Name – *Chaetodon striatus*

Local Name – Swede, Butterflyfish

Visual ID – White with two black bands on body and a further band across the eye. The rear of body, together with the border of dorsal, anal and caudal fins, is also black. Juvenile is similar but has an ocellated spot on the rear of its dorsal fin. Maximum size c.15 cm.

Abundance, Distribution & Behaviour - Occasional, although can be locally common. Seen on most reef habitats around Anguilla. They swim about the reef, largely unconcerned. Adults are often seen in pairs.

Common Name – **Spotfin Butterflyfish**

Scientific Name – *Chaetodon ocellatus*

Local Name – Swede, Butterflyfish

Visual ID – White with yellow on all peripheral fins. Black bar passes through eye and rear outer edge of pectoral fin has small black spot. Juvenile similar, but with black bar on rear body and translucent caudal fin. Maximum size c.20 cm.

Abundance, Distribution & Behaviour – Uncommon. Has been sighted at Anguillita and other more offshore reef habitats. Swim about the reef largely unconcerned. Adults are often seen in pairs.

Common Name – **Reef Butterflyfish**

Scientific Name – *Chaetodon sedentarius*

Local Name – Swede, Butterflyfish

Visual ID – White with yellowish back, dorsal and caudal fin. Black rear of body, dorsal and anal fins. Black bar passes across the eyes. Juveniles are very similar but may have a small black spot on the rear of the dorsal fin. Maximum size c.15 cm.

Abundance, Distribution & Behaviour - Occasional in some areas, uncommon in others. Can be seen on most reef habitats, although is most likely seen on offshore reefs. Swim about the reef largely unconcerned. Adults are often seen in pairs.

Common Name – **Longsnout Butterflyfish**

Scientific Name – *Chaetodon aculeatus*

Local Name – Swede, Butterflyfish

Visual ID – White with yellowish upper body which darkens with age and long pointed snout. Dorsal spines are mohawk-like when pushed forward. Has dusky to yellow bar running from the upper forehead to the eye. Translucent caudal fin. Maximum size c.10 cm.

Abundance, Distribution & Behaviour - Uncommon in most areas, it generally inhabits deeper reefs and walls making sightings while snorkelling rare. Unlike other species, the Longsnout Butterflyfish is usually solitary and secretive, darting into dark reef recesses when approached.

Common Name – **Foureye Butterflyfish**

Scientific Name – *Chaetodon capistratus*

Local Name – Swede, Butterflyfish

Visual ID – White with yellowish tinge to their underside. Has an ocellated spot on rear of body and often a faded bar through the eye. Numerous thin dark lines radiate diagonally from mid-body. Juveniles (below) have second smaller ocellated spot above the larger one, and two dusky body bars which fade with age. Maximum size c.15 cm.

Abundance, Distribution & Behaviour – Common. Can be seen on most reef habitats, with juveniles also common on seagrass beds. Swim about the reef largely unconcerned. Adults are often seen in pairs.

Family – Acanthuridae: Surgeonfishes

Surgeonfishes, one of the most common fish Families within Anguilla's reef habitats, derive their name from a spine, as sharp as a surgeon's scalpel, found on either side of their caudal peduncle. They use these spines, which are usually held flat but hinged from the back end, as a means of defence by flapping their tails from side to side; consequently fishers pick them up with caution! Surgeonfish are herbivores and can often be seen swimming in loose aggregations feeding off algae.

Common Name – **Blue Tang**

Scientific Name – *Acanthurus coeruleus*

Local Name – Ocean Doctor, Blue Doctor

Visual ID – Blue body with a steep sloping forehead and white spine on caudal peduncle. Juveniles are brilliant yellow (bottom), and intermediate phases may be blue with a yellow caudal fin. At night the adult is often dark with white smudges. White spine distinguishes the Blue Tang from other surgeonfish species. Maximum size c.35 cm.

Abundance, Distribution & Behaviour - Abundant. Present on all reef habitats in Anguillian waters. Can be seen foraging for algae on their own or in large aggregations with other surgeonfish species. Generally, they do not seem concerned by sightseers although will swim off if closely approached. A few fishers target this species in Anguilla by baiting their traps with what is locally known as sea moss.

Common Name - **Ocean Surgeonfish**

Scientific Name – *Acanthurus bahianus*

Local Name – Ocean Doctor, Brown Doctor

Visual ID – Grey to dark brown body with gently sloping forehead body and dark spine on caudal peduncle. Very similar to the Doctor Fish (next) but does not have body bars and its pectoral fin is transparent, sometimes with a yellow tinge. It also has a more crescent-shaped caudal fin than the Doctor Fish. Can change colour and sometimes has a white band on its caudal peduncle. Juveniles are miniature versions of the adult. Maximum size c.35 cm.

Abundance, Distribution & Behaviour - Abundant. Present on all reef habitats in Anguillian waters. Can be seen foraging for algae on their own or in large aggregations with other surgeonfish species. Generally, they do not seem concerned by sightseers although will swim off if closely approached. Juveniles are often seen on seagrass beds.

Common Name – **Doctor Fish**

Scientific Name – *Acanthurus chirurgus*

Local Name – Ocean Doctor, Fan Tail

Visual ID – Grey to dark brown body with gently sloping forehead and dark spine on caudal peduncle. Very similar to the Ocean Surgeonfish (previous) but has body bars and a dark-coloured pectoral fin. It also has a less crescent-shaped caudal fin than the Ocean Surgeonfish. Can change colour and sometimes has a white band on its caudal peduncle. Juveniles are miniature versions of the adult. Maximum size c.35 cm.

Abundance, Distribution & Behaviour – Common to occasional. The least common surgeonfish but is still an easy reef resident to spot once confusion with the Ocean Surgeonfish is avoided. Can be seen foraging for algae on their own or in large aggregations with other surgeonfish species. Generally, they do not seem concerned by sightseers although will swim off if closely approached. Juveniles are often seen on seagrass beds.

Family – Pomacentridae: Damselfishes & Chromis

Damselfishes are energetic little creatures and a characteristic part of the reef community. They are oval-shaped and, as adults, often plain in colour. Juveniles are usually more colourful. Several species spend their day busily tending an algal garden which they defend vigorously. They don't actually feed off this filamentous 'turf' algae (page 14), but rather on epiphytes which grow on it and other associated detritus. These gardeners will chase off large herbivorous parrotfish, sometimes high into the water column, and will even nibble at divers. The Sergeant Major and Bicolor Damselfish are more similar to chromis, feeding on plankton in the water column; they are therefore non-aggressive, unless defending eggs. They also differ from the other damselfishes, in a similar way to the chromis, by having juveniles that are almost identical miniature versions of the adult. Interestingly, the Bicolor Damselfish is probably the most sophisticated vocalist on Anguillian reefs; they use a complicated repertoire of calls, in combination with other displays, colour changes and chemical signals, to communicate with other members of their size-related hierarchies.

Some species of damselfish are difficult to differentiate from one another, for example the dark bodied Longfin, Dusky, Beaugregory, Cocoa (not featured) and Threespot, although subtle differences usually make underwater identification possible.

Common Name – **Longfin Damselfish**

Scientific Name – *Stegastes diencaeus*

Local Name – Katie, Co-Pilot

Visual ID – Dark with elongated dorsal and anal fins that may be somewhat pointed and extend well beyond caudal peduncle. Adult (top) closely resembles the Dusky Damselfish (next). The Longfin Damselfish juvenile (bottom right) has brilliant yellow-gold body and obvious blue speckles that form clear stripes down the head towards an ocellated spot on rear dorsal fin. When transitioning into an adult, the juvenile's markings and colourations gradually fade (bottom left) Maximum size c.12 cm.

Abundance, Distribution & Behaviour – Common, although confusion between the Dusky Damselfish means precise abundances are difficult to assess. Numbers also vary greatly from location to location. Can be seen on most reef habitats and is unafraid, aggressively defending its algal garden territory.

Photo Credit Florent Charpin

Photo Credit Gert-Jan Vandenbos

Photo Credit Lippold Hakken

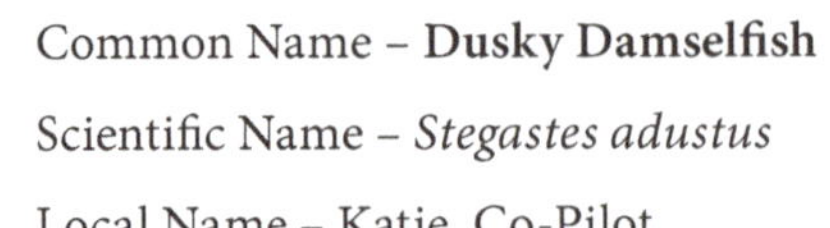
Photo Credit Florent Charpin

Photo Credit Steve Donahue

Common Name – **Dusky Damselfish**

Scientific Name – *Stegastes adustus*

Local Name – Katie, Co-Pilot

Visual ID – Dark with rounded dorsal and anal fins which rarely extend beyond the caudal peduncle. Adult (top) closely resembles the Longfin Damselfish (previous). Juvenile (bottom) has brilliant, sunset-orange, dusky wash over head and fore-back. The juvenile also has blue speckles on forehead with small black blotch on caudal peduncle and ocellated spot on rear dorsal fin. Maximum size c.15 cm.

Abundance, Distribution & Behaviour – Occasional, with juveniles more frequently sighted than adults; this may be due, in part, to confusion between adult Longfin Damselfish. Can be seen on most reef habitats. Unafraid and probably the most aggressive territorial damselfish, especially when defending eggs.

Photo Credit Stuart Wynne

Photo Credit Stuart Wynne

Common Name – **Threespot Damselfish**

Scientific Name – *Stegastes planifrons*

Local Name – Katie, Co-Pilot

Visual ID – Adults (top) are yellowish-brown to grey, with dark spot at base of each pectoral fin, and another on the caudal peduncle. Juveniles (bottom) are bright yellow with dark spot on caudal peduncle and another on the dorsal fin. Very young Rock Beauties (page 133) can appear somewhat similar. Maximum size c.12 cm.

Abundance, Distribution & Behaviour – Occasional, although can be locally common, for example in Shoal Bay East. Prefers reef tops and can sometimes be seen swimming in loose aggregations. Territorially defends its algal garden with vigour.

Common Name – **Beaugregory**

Scientific Name – *Stegastes leucostictus*

Local Name – Katie, Co-Pilot

Visual ID – Bluish-grey body with yellowish to white belly and caudal fin. The Beaugregory juvenile has a yellow body with a blue wash over forehead similar to Longfin Damselfish juvenile (page 139), but lacks the ocellated spot, instead having a small dark blotch on dorsal fin, and sometimes a black spot on the caudal peduncle. Blue speckles on forehead do not usually form obvious lines. The pictured individual is in the process of transitioning between juvenile and adult. Maximum size c.10 cm.

Abundance, Distribution & Behaviour – Common, especially in some reef habitats where they may be the most common species of damselfish present. It lives in a variety of habitats including seagrass beds, sand, rubble and rocky reefs, but seems to prefer reef tops. Not as aggressive as other damselfish although is still territorial, especially if guarding eggs.

Photo Credit Gert-Jan Vandenbos

Common Name – **Yellowtail Damselfish**

Scientific Name – *Microspathodon chrysurus*

Local Name – Katie

Visual ID – Dark, bulky body with heavy lips. Normally has yellow caudal fin which, on occasion, is black. Adults can resemble the Yellowtail Hamlet (page 145) although are noticeably bulkier, especially when older (top). Sometimes has some metallic blue speckles on head and back. Juvenile (bottom) is dark blue with metallic spots all over its body; often being referred to as a 'Jewelfish'. It has a transparent tail which begins to turn yellow when still quite young. Maximum size c.20 cm.

Abundance, Distribution & Behaviour – Common. Usually inhabits reef tops and maintains a small territory. It is not aggressive and is unafraid, swimming about the reef in and out of recesses patrolling its domain.

Photo Credit Florent Charpin

Photo Credit Steve Donahue

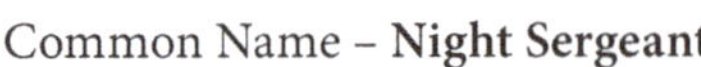

Common Name – **Night Sergeant**

Scientific Name – *Abudefduf taurus*

Local Name – Katie

Visual ID – Grey-Brown with five dark body bars. Very similar to the Sergeant Major although its top lip overhangs the lower. Face has a dark wash and it usually holds its dorsal fin more aloft than the Sergeant Major. Maximum size c.25 cm.

Abundance, Distribution & Behaviour – Uncommon. Prefers shallow rocky inshore areas, especially wave-cut shelves and overhangs. Has been sighted between Pelican Point and Little Bay and also along the sea rocks at Maundays Bay. They become more active at night but their nervousness means they are generally hard to observe.

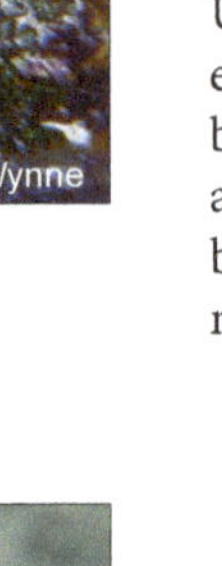

Common Name – **Sergeant Major**

Scientific Name – *Abudefduf saxatilis*

Local Name – Katie

Visual ID – Silver with yellow upper body and five distinct black bars. Does not have an overhanging top lip like the very similar Night Sergeant (previous). Maximum size c.15 cm.

Abundance, Distribution & Behaviour – Common although can be locally abundant. Swims above most reef habitats, usually mid-water and in aggregations. When guarding eggs the males darken dramatically and swim in amongst the rocks in a very agitated fashion.

Common Name – **Bicolor Damselfish**

Scientific Name – *Stegastes partitus*

Local Name – Katie, Co-Pilot

Visual ID – Fore-body is usually black with white back-body, often having yellow area on belly and mid-body. Colourations can be highly variable but this is a very distinctive species. Maximum size c.10 cm.

Abundance, Distribution & Behaviour – Common, although may be locally abundant inhabiting patch reefs and low-profile reef tops. Is often seen over small sand patches and rubble areas, but equally favours low profile reefs where it hovers patiently over a sponge or protective recess that it has chosen to call home. Non-aggressive, apart from territorial egg-guarding males.

Common Name – **Blue Chromis**

Scientific Name – *Chromis cyanea*

Local Name – Blue Chromis

Visual ID – Metallic blue with a dark-bordered, deeply forked tail. Upper body has a dark margin sometimes all the way to the forehead. Similar to the juvenile Creole Wrasse (page 162) and, at times, the two species have been observed swimming together. Males change colour dramatically during courtship, with their bellies becoming almost silver. Females turn black while laying eggs. Maximum size c.12 cm.

Abundance, Distribution & Behaviour – Common to abundant. Found swimming above most reef habitats, usually quite close to the reef surface. Can form large schools, especially above large coral heads where they might mix with Brown Chromis (next).

Common Name – **Brown Chromis**

Scientific Name – *Chromis multilineata*

Local Name – Brown Chromis

Visual ID – Brownish grey to silver with yellow borders to its dorsal caudal fins. Has a black base to its pectoral fin. Maximum size c.15 cm.

Abundance, Distribution & Behaviour – Common, but may be locally abundant. Can form very large schools above the reef spreading out into open-water. Smaller individuals sometimes school over coral heads and mix with Blue Chromis (previous).

Common Name – **Sunshinefish**

Scientific Name – *Chromis insolata*

Local Name – Sunshinefish

Visual ID – Adult has dark body with yellow to transparent rear dorsal fin and caudal fin margin. Juvenile (pictured) has brilliant yellow upper body and purple-blue underside. Maximum size c.10 cm.

Abundance, Distribution & Behaviour – Rare. However, as it usually prefers deeper reefs actual abundances are currently unknown.

Large Lips and Heavy Bodies

Family – Serranidae: Seabasses (Part 1: Hamlets)

Hamlets, although similar in appearance to damselfishes, are easily distinguished by their flat head, obvious ventral fins and large lips – the latter also being a clue that they belong to the seabass Family. They are usually colourful but there has been a long-running debate amongst Caribbean ichthyologists as to whether they all belong to one species (*Hypoplectrus unicolor*) because different colour morphs can mate together, or whether they are, in fact, many different species that are simply able to hybridise. As all juvenile hamlets look the same and because some colouration patterns are very similar (see head markings on the Barred Hamlet and Yellowbelly Hamlet later in this section as an example), many people still believe that all hamlets belong to one species with diverse colourations. However, the American Fisheries Society now considers the common distinct colour patterns to be different species. Lab-based experiments led to this decision, when it was observed that the vast majority of hamlets favour mates of the same colour and pattern, only hybridising if there are no similar individuals present. This debate is still underway and is especially interesting because, at its heart, is the tricky question, 'What defines a species?'. The answer may never be agreed upon by everyone.

Hamlets are also interesting because they reproduce by a means known as egg-trading. Like some of the other small seabasses they are known as simultaneous hermaphrodites, which means that during courtship the role of the male and female are assumed alternately by both individuals, thus sharing the energetic burden of reproduction. As bizarre as this sounds, producing eggs is far more costly in terms of energy than producing sperm. By alternating roles the number of fertilised eggs can be maximised, out-doing total production when fish use other methods of reproduction. Hamlets only live in the tropical Atlantic and, being carnivorous, feed off small fish and crustaceans. The most diverse population of hamlets in Anguillian waters is found in Shoal Bay East where almost all the species listed here can be seen regularly, together with a variety of hybrids.

Common Name – **Juvenile Hamlets**

Scientific Name – n/a

Local Name – None

Visual ID – White to pale gold with obvious black and white markings on caudal peduncle and faint facial patterns. Species colourations become apparent as maturity is reached, which happens at c.5 cm.

Abundance, Distribution & Behaviour – Occasional, although can be locally common or abundant forming schools around coral heads. Also sometimes sighted on seagrass beds and sandy bottoms.

Common Name – **Butter Hamlet**

Scientific Name – *Hypoplectrus unicolor*

Local Name – Butter Hamlet

Visual ID – White with large black saddle on caudal peduncle. Has bright blue markings on head and a black blotch in front of the eye. Maximum size c.12 cm.

Abundance, Distribution & Behaviour – Uncommon, but is considered occasional in Shoal Bay East. Remains near bottom, has small reef territory.

Common Name – **Barred Hamlet**

Scientific Name – *Hypoplectrus puella*

Local Name – Barred Hamlet

Visual ID – Brown body with white bars. Has bright blue vertical lines on head and around eye with spots on snout. Maximum size c.15 cm.

Abundance, Distribution & Behaviour – Occasional, although can be locally common and is present on most reef habitats around the island. Remains near bottom, has small reef territory.

Common Name – **Yellowbelly Hamlet**

Scientific Name – *Hypoplectrus aberrans*

Local Name – None

Visual ID – Yellow underside and caudal fin, with dark wash over back. Has blue markings on head and around eye, with spots on snout. Maximum size c.12 cm

Abundance, Distribution & Behaviour – Uncommon. Remains near bottom, has small reef territory.

Common Name – **Yellowtail Hamlet**

Scientific Name – *Hypoplectrus chlorurus*

Local Name – None

Visual ID – Dark body with yellow caudal fin. Not to be confused with the similar Yellowtail Damselfish (page 141). Maximum size c.12 cm.

Abundance, Distribution & Behaviour – Uncommon to occasional. Remains near bottom, has small reef territory.

Common Name – **Black Hamlet**

Scientific Name – *Hypoplectrus nigricans*

Local Name – None

Visual ID – Dark coloured, with long ventral fins that are usually pointed downwards. Maximum size c.15 cm.

Abundance, Distribution & Behaviour – Occasional although can be locally common, for example at Pelican Point. Remains near bottom, has small reef territory.

Common Name – **Hybrid Hamlets**

Scientific Name – n/a

Local Name – None

Visual ID – Varies greatly, with the parentage often able to be guessed by colour patterns. The two examples pictured below are probably hybrids of the Black and the Yellowtail Hamlet (below left), and the Barred and the Yellowbelly Hamelt (below right).

Abundance, Distribution & Behaviour – Uncommon to occasional. Some areas seem to have a hamlet population of hybrids only, especially noted in some parts of Shoal Bay East.

Family – Serranidae: Seabasses (Part 2: Groupers)

Groupers are the best known members of the seabass Family, with strong plump bodies, large mouths and heavy lips. They are solitary hunters that spend most of their time lurking in reef shadows, wrecks or under ledges and rocks. They remain stationary for extended periods, and can change colours rapidly, allowing them to blend well with the background. This colour change can make some grouper species hard to differentiate at first. They catch prey by rapidly opening their large mouths, creating a powerful suction that draws water and any unsuspecting fish or crustaceans into their gullets. Groupers are hermaphroditic, beginning life as a female and changing into a male with maturity.

Groupers were once extremely common members of the reef community but over recent decades, due to fishing pressure, numbers have become greatly reduced. The reason they have succumbed to this pressure more severely than other reef fish is not just because they are favoured by fishers; when breeding most grouper species form enormous transient spawning aggregations in specific locations, which historically consisted of thousands of individuals. Gradually, as fishers learnt about the locations and specific seasonality of such events, they could, in a very short space of time, remove the majority of mature groupers from whole areas.

Common Name – **Goliath Grouper**

Scientific Name – *Epinephelus itajara*

Local Name – Grouper

Visual ID – Mottled yellowish brown to olive green body with small dark spots. Juveniles may have body bars present. Is the largest reef-dwelling fish species in Anguillian waters. Maximum size c.2 m.

Abundance, Distribution & Behaviour – Rare. Reclusive, usually hiding in caves, wrecks or under ledges. One individual has been sighted on a number of occasions at Dog Island, and other reports of one at Anguillita were made a few years ago. More recently sightings have been made on a number of the wrecks, which are suspected to be of the same individual who moves between locations. Estimations are that it weighs more than 50 kg (100-120 lbs). This species was formally known by the common name 'Jewfish'.

Common Name – **Nassau Grouper**

Scientific Name – *Epinephelus striatus*

Local Name – Nassau Grouper

Visual ID – Five irregular olive-brown bars over light background. Has black saddle on caudal peduncle and can change body colour dramatically to become almost white or black. Maximum size c.1 m.

Abundance, Distribution & Behaviour – Uncommon to occasional. Can be found on most reef habitats although has probably suffered from harvesting more than any other grouper species in Anguilla. Juveniles are sometimes seen on seagrass beds in amongst Turtle Grass blades (page 15) see below. Adults spend much of their time lying on the bottom blending in with their surroundings.

Photo Credit Stuart Wynne

Photo Credit Stuart Wynne

Photo Credit Lippold Hakken

147

Common Name – **Graysby**

Scientific Name – *Cephalopholis cruentatus*

Local Name – Graysby

Visual ID – Light red to grey body covered with orangish spots. Has three to five dark spots along base of dorsal fin. These spots are sometimes white, especially in juveniles. Can lighten and darken dramatically. Caudal fin highly rounded. Maximum size c.30 cm.

Abundance, Distribution & Behaviour – Occasional to common, sometimes being the most abundant grouper species present on a reef. It has a nervous disposition compared to other grouper species and it often retreats for cover before being spotted by divers; unsurprisingly, it prefers reefs with more hiding places.

Common Name – **Red Hind**

Scientific Name – *Epinephelus guttatus*

Local Name – Red Hind

Visual ID – Whitish body with reddish spots which are usually larger than those seen on the Graysby (previous), but more dense than those on the Rock Hind (next). Dorsal, anal and caudal fins have white-edged black margin. Can lighten and darken dramatically. Maximum size c.60 cm.

Abundance, Distribution & Behaviour – Occasional, inhabiting all reef areas from shallow inshore patch reefs to deeper banks. Drifts just above the bottom or rests on pectoral fins, and is arguably the most inquisitive grouper species.

Common Name – **Rock Hind**

Scientific Name – *Epinephelus adscensionis*

Local Name – Rock Hind

Visual ID - Whitish body with reddish spots which are usually larger than those seen on the Graysby (top of page), but less dense than those on the Red Hind (previous). Has a dark saddle on the caudal peduncle, and one to four pale or dark blotches along the back. Can lighten or darken dramatically. Maximum size c.60 cm.

Abundance, Distribution & Behaviour – Uncommon, but may inhabit most reef areas and seems to favour shallow rocky, often rough, inshore areas. Drifts near the bottom blending with background.

Common Name – **Coney**

Scientific Name – *Cephalopholis fulva*

Local Name – Butterfish

Visual ID – Body colours are highly variable, from deep red (common) to golden (rare). May be bicoloured with a red back and white belly. Body is covered with small blue dots which are more obvious in paler phases. Key identification features are two black dots on lower lip and two black dots behind dorsal fin on caudal peduncle. Juveniles are usually golden. Maximum size c.40 cm.

Abundance, Distribution & Behaviour – Common on most reef habitats, usually the most regularly seen grouper species. Drifts near the bottom but often seen actively swimming about the reef. Is sometimes gregarious forming small groups. Golden juveniles have been sighted around Anguillita, and golden adults offshore from Prickly Pear.

Photo Credit Lippold Hakken

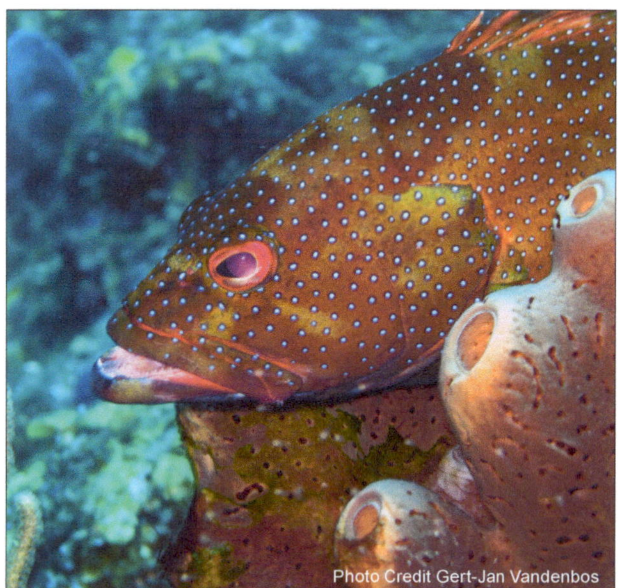
Photo Credit Gert-Jan Vandenbos

Photo Credit Lippold Hakken

Common Name – **Black Grouper**

Scientific Name – *Mycteroperca bonaci*

Local Name – Grouper

Visual ID – Reddish brown to black with thin pale yellow margin on pectoral fins. Although very similar to Yellowfin Grouper (page 150), the ends of the body blotches, which are almost square, allow reliable identification. However, these blotches become indistinct while paling or darkening. When pale, thick black margins of rear fins become obvious. Maximum size c.1 m.

Abundance, Distribution & Behaviour – Uncommon, drifting near bottom of reef. Are reported to be more common in shallow offshore fishing areas such as Old England. Have also been sighted at Sombrero Island.

Photo Credit Lippold Hakken

Photo Credit Stuart Wynne

Photo Credit Lippold Hakken

Common Name – **Tiger Grouper**

Scientific Name – *Mycteroperca tigris*

Local Name – Grouper

Visual ID – Nine brown to black diagonal tiger stripes over light background. Can dramatically change colour, pale or darken. Tiger stripes fade into blotched pattern on belly and head. Juveniles begin life golden with dusky mid-body streak (inset image). Adult pattern gradually appears and colour changes as maturity is reached. Maximum size c.1 m.

Abundance, Distribution & Behaviour – Uncommon to rare, inhabiting reefs and walls. Sometimes rests or drifts over bottom but often actively swims around the reef. Regularly frequents cleaning stations.

Photo Credit Stuart Wynne

Common Name – **Yellowfin Grouper**

Scientific Name – *Mycteroperca venenosa*

Local Name – Gag

Visual ID – Black to red or pale olive-green speckles over light background, with irregular rectangular blotches that may be indistinctive as can lighten and darken dramatically. The ends of the body blotches are rounded and rear fins have thin dark margins, differentiating it from the Black Grouper (page 149). Outer third of pectoral fin is pale to bright yellow. Maximum size c.1 m.

Abundance, Distribution & Behaviour – Uncommon to rare, inhabiting reefs and walls. Sometimes rests or drifts over bottom but also commonly seen swimming about.

Photo Credit Stuart Wynne

Common Name – **Red Grouper**

Scientific Name – *Epinephelus morio*

Local Name – Gag

Visual ID – Usually reddish brown with scattered white blotches, patches and bars, sometimes resembling the Nassau Grouper (page 147). Can lighten and darken dramatically. Maximum size c.1 m.

Abundance, Distribution & Behaviour – Uncommon to rare. The pictured individual is the only one recorded in Anguillian waters at the time of writing, seen on the wreck of the M.V. Oosterdiep.

Family – Serranidae: Seabasses (Part 3: Others)

Often called bass, these seabass Family members are often more colourful and usually much smaller than the groupers. Their bodies are also not as bulky, being more cylindrical but, like the hamlets, they still possess relatively heavy lips and large mouths. Most stay near the bottom or inhabit dark recess, except for the Creolefish which is often in open-water. The soapfish were formerly classified as a separate Family (Grammistinae), but are now grouped together with the seabass.

Common Name – **Harlequin Bass**

Scientific Name – *Serranus tigrinus*

Local Name – None

Visual ID – Whitish upper body and yellowish belly, with dark tiger stripes and mottled dusky colourations that form subtle horizontal stripes along the body. Fins and head also have scatterings of dusky spots and yellow highlights. Maximum size c.10 cm.

Abundance, Distribution & Behaviour – Common on most reef habitats to those with a keen eye. Favours low profile areas, coral rubble and also seagrass beds. They are often seen in pairs hunting for small crustaceans, but when motionless can be hard to spot as they blend very effectively with their background.

Common Name – **Lantern Bass**

Scientific Name – *Serranus baldwini*

Local Name – None

Visual ID – Pale body with reddish-brown to orange rectangular blotches and spots which are especially obvious on belly. Has black spots on caudal fin and a yellow stripe along its frontal mid-body. Maximum size c.7 cm.

Abundance, Distribution & Behaviour – Uncommon. Seen in Little Bay hiding amongst Turtle Grass blades (page 15) but also favours coral rubble. Can be found in water beyond safe-diving limits when, reportedly, individuals are more reddish-orange or yellow than those from shallower parts.

Do groupers bite? Although most species of grouper are sullen, timid creatures who disappear into a recess if approached, they have a sharp set of teeth that could inflict a nasty injury to an unsuspecting finger (see inset top image on page 144). This is especially the case for larger individuals, who unsurprisingly also have larger teeth! Groupers are not vigorously territorial but will at times defend a favoured hidey-hole, especially if they have claimed the perfect retreat in a sunken wreck or cave. These larger species have been known to snap warning bites and act defensively towards divers who approach too enthusiastically, although no reports of actual attacks are known. Not yet anyway!

Common Name – **Tobaccofish**

Scientific Name – *Serranus tabacarius*

Local Name – None

Visual ID – Mid-body is shades of orange to brown with alternating series of dark and bright white blotches along back. Caudal fin has dark U-shape pattern, belly is white. Maximum size c.15 cm.

Abundance, Distribution & Behaviour – Locally common, inhabiting reefs on adjacent areas of sand and coral rubble. They tend to stay near the bottom and are often seen in pairs. Limestone Bay is a good location to spot this species, as it is present on almost all small rubble patches in the vicinity of the dive mooring.

Common Name – **Chalk Bass**

Scientific Name – *Serranus tortugarum*

Local Name – None

Visual ID – Bluish white with series of pale to dark bars along back. Occasionally, the body can be darker. Maximum size c.10 cm.

Abundance, Distribution & Behaviour – Uncommon, although can be seen hovering in small groups over sand and coral rubble within deeper reef areas. Has been sighted close to the wreck of the M.V. Oosterdiep and on sand flats beyond Limestone Bay reef.

Common Name – **Creolefish**

Scientific Name – *Paranthias furcifer*

Local Name – Snapper

Visual ID – Darkish-blue/silver to reddish brown. Has bright red spot at base of pectoral fin and three light or dark spots along back. Forked caudal fin. Maximum size c.25 cm.

Abundance, Distribution & Behaviour – Uncommon, although can be locally abundant on deeper offshore reefs where large schools may form. Feed on plankton, hence their preference for areas where water generally circulates more freely. A common species to see on some of the wrecks and around Dog Island. Note Isopod on cheek (page 83).

Common Name – **Greater Soapfish**

Scientific Name – *Rypticus saponaceus*

Local Name – Soapfish

Visual ID – Drab reddish-brown to grey, with mottled pattern formed by pale, indistinct blotches. Has pointed head with protruding jaw and rounded caudal fin. A covering of mucus produces soapsud-like bubbles when caught or handled, hence their common name. Maximum size c.35 cm.

Abundance, Distribution & Behaviour – Uncommon to occasional, depending on location. May be present on most reef habitats although their reclusive nature means they are rarely seen. They generally inhabit shallower waters and are solitary night feeders. During the day they lie on the bottom, leaning against the back of a protective overhang or recess. Juveniles are usually more active and can commonly be seen at sites around Dog Island. Reportedly adults are often caught by fishers in Sandy Hill Bay. Although their mucus does not deter cleaner fish at cleaning stations, it does contain a protein toxin, and soap fish have yet to be discovered in the belly of a predator.

Family – Grammatidae: Basslets

The basslet Family is closely related to seabass but all are small and lack a continuous lateral line. Most members of the Family inhabit deeper waters and walls (for example, the Blackcap Basslet – *Gramma melacara*), so are not normally seen when diving or snorkelling in Anguillian waters. Fairy Basslets are an exception to this but are still often overlooked as they inhabit the underside of undercuts and recesses in the reef.

Common Name – **Fairy Basslet**

Scientific Name – *Gramma loreto*

Local Name – None

Visual ID – Bicoloured body with purple front half and yellow rear. Easily confused with juvenile Spanish Hogfish (page 161), although a dark spot on their fore-dorsal fin is a good clarification mark. Although small they have been reported to grow up to a maximum size of c.7 cm.

Abundance, Distribution & Behaviour – Occasional, although can be locally common if suitable habitat is present. They live upside-down, almost exclusively inhabiting the underside of undercuts, caves and other reef recesses. This behaviour easily distinguishes them from the similar juvenile Spanish Hogfish.

Swim with Pectoral Fins

Family – Scaridae: Parrotfishes

Parrotfishes, who are among the largest and most common reef-dwelling species, are characterised by powerful, beak-like jaws and fused teeth. This beak, combined with their bright colour patterns, is where their common name derives. They are closely related to wrasses, but have bulkier bodies and are generally a lot larger in size. Wrasses also have noticeable teeth, and swim by continuously flapping their pectoral fins, whereas parrotfishes flap them a few a times and glide for a short distance before repeating the action.

Parrotfishes often form harems (see second paragraph following) but do not usually school, although juveniles may join together in loose aggregations of mixed species. Gatherings also occur during resident spawning aggregations and as large groups of males who rapidly cruise the reef. This latter unexplained phenomenon has been observed in a number of locations in Anguillian waters, and usually consists of male Spotlight Parrotfish with a few Redband Parrotfish tagging along.

Although herbivores, parrotfishes can often be seen taking large bites out of hard coral, and may seem to focus their attention in one particular location, sometimes causing considerable damage to a small area. The reason for this behaviour, known as 'focus biting', is not known but it is believed that they bite the coral and swallow the hard skeleton to aid digestion, also gaining nutritional rewards not only from the coral polyps, but also from its symbiotic algae, known as zooxanthellae. It is a fascinating thought that parrotfish excretions are one of the primary sources of the sand found on tropical beaches, together with certain species of green algae (page 19-24).

Like wrasses, as parrotfishes grow up they usually first develop and mature as a female, living in a harem with a dominant male. Being sequential protogynous hermaphrodites they have the ability to change from female to male as maturity is reached. As a general rule the harem's largest female will develop into a dominant male once the existing lead male is no longer present. Generally these developmental 'phases' are split into juvenile phase (JP), initial phase (IP) and terminal phase (TP). Terminal phase are sexually mature males; initial phase includes sexually mature females and, in some species, immature and mature males; juvenile phase, as its name suggests, contains juvenile individuals. These different 'phases' are very distinct from each other, unlike some other sex-changing fish Families, for example, the groupers (page 146-150). Occasionally there may be an additional intermediate colour phase between the three primary phases, and some even change colour rapidly, making identification even more of a challenge.

Interestingly, all parrotfishes in the Caribbean are believed to have evolved from two wrasse-like members of the Family, the rare Emerald Parrotfish (not featured in this book), and the slightly more common Bluelip Parrotfish (page 160). Both species belong to distinct Genera, and have an elongated body and no obvious beak, thus forming a link between the two somewhat similar Families.

How important are parrotfish? As mentioned earlier, parrotfish are considered significant grazers of algae, and also contribute a lot of material that helps construct the beautiful beaches that we all know and love throughout the Caribbean. This role is considered so crucial to the coral reef and surrounding ecosystems that parrotfish are often referred to as keystone species and in many places are protected by law. Over recent years however, some research-ers speculate that, due to overall reef degradation and the habit parrotfish have of focus biting, they may today be causing detriment to the reef and make it harder for new corals to settle and recolonise: with fewer corals populating the reef, the effect of focus biting is more noticeable and those corals that do settle and grow are more precious. In reality, although there may be some truth to this, the role parrotfish play as grazers probably out-weights the dam-age they do, and it is far better to try and keep the fish community intact in the hope that one day coral populations return to past levels, rather than manage a diminishing coral population by reducing other natural community members.

Common Name – **Rainbow Parrotfish**

Scientific Name – *Scarus guacamaia*

Local Name – Blue Witch

Visual ID – TP: Orange to brown head and caudal fin with green midbody and slightly humped back (top). Caudal fin has relatively long trailing edges. IP & JP: Generally green although scale edges are orange to brown. The largest species of parrotfish in Caribbean waters that can grow to over 1.5 m.

Abundance, Distribution & Behaviour – Rare. Potentially this species could be found on any reef habitat but, over recent years, has only been sighted in a couple of locations (including Sile Bay). This suggests overfishing. Swims about the reef scraping algae from rocks and corals.

Common Name – **Queen Parrotfish**

Scientific Name – *Scarus vetula*

Local Name – Parrotfish (TP), Mule (IP)

Visual ID – TP: Green with dramatic blue and dark green markings around the mouth (top). Has a light bar on the pectoral fin and the caudal fin has bold edges with trailing tips. IP & JP: Dark grey to brownish-black body. A broad white stripe with diffused edges runs down the mid-body (bottom). Maximum size c.60 cm

Abundance, Distribution & Behaviour – Common to occasional on most reef habitats, although can be somewhat variable between locations. Swims about the reef scraping algae from rocks and corals.

Photo Credit Florent Charpin

Photo Credit Lippold Hakken

Photo Stuart Wynne

Common Name – **Stoplight Parrotfish**

Scientific Name – *Sparisoma viride*

Local Name – Specktail (TP), Redbelly (IP)

Visual ID – TP: Emerald green with salmon-coloured markings on head and fins. Has bright yellow spot at upper corner of gill cover and yellow area at the base of its caudal fin with yellow to salmon crescent behind (top). Resembles the TP Redband Parrotfish (page 157) but generally looks bulkier. IP: Mottled red-brown and white upper body formed from scale variations, with a red belly and caudal fin (bottom left). JP: Reddish brown with three rows of widely spaced white spots and white patch on caudal fin (bottom right). Maximum size c.60 cm.

Abundance, Distribution & Behaviour – Common on all reef habitats where it swims about the area stopping to scrape algae from rocks and coral.

Photo Credit Stuart Wynne

Photo Credit Stuart Wynne

Common Name – **Princess Parrotfish**

Scientific Name – *Scarus taeniopterus*

Local Name – Parrotfish

Visual ID – TP: Blue to green with gold blotch or stripe behind the pectoral fin fading towards rear. Two blue to green stripes pass across eye, and caudal fin has pinkish border (top). Resembles TP Striped Parrotfish (next) but is noticeably larger with a bulkier body. IP: Goldish with yellow fins, dark border to the caudal fin, and visibly faded stripes. JP: Dark with white stripes, one of which is directly under the dorsal fin, and dark borders to the caudal fin (bottom). Both these younger phases are also very similar to the Striped Parrotfish. Maximum size c.35 cm.

Abundance, Distribution & Behaviour – Occasional, although can be seen on most reef habitats where it swims about the area stopping to scrape algae from rocks and coral.

Common Name – **Striped Parrotfish**

Scientific Name – *Scarus iserti*

Local Name – Parrotfish

Visual ID – TP: Blue to green with gold to yellow stripe behind the pectoral fin, and paler one below (top). Resembles the Princess Parrotfish (previous) but yellow colour is less dramatic and body less bulky. IP & JP: Also very similar to the Princess Parrotfish although they usually have an obvious yellow nose and no dark border to caudal fin (bottom left & right). Maximum size c.25 cm.

Abundance, Distribution & Behaviour – Common to abundant, especially JP, and can be seen in small to large groups in certain areas. The most common parrotfish. Swims about the area stopping to scrape algae from rocks and coral.

Common Name – **Redband Parrotfish**

Scientific Name – *Sparisoma aurofrenatum*

Local Name – Parrotfish (TP), Tomteg (IP)

Visual ID – TP: Green body. Has small yellow blotch with two or more small black spots on upper body and red band from corner of mouth to under eye (top). Resembles Spotlight Parrotfish (page 156) but has red borders to caudal fin, red anal fin, and is generally smaller. IP: Highly variable and changes colour rapidly, sometimes displaying a striped pattern, although most commonly has an olive-green mottled body with red fins (bottom left). It can pale dramatically and look very similar to the IP Redtail Parrotfish (next); most of the time it is clearly distinguishable by a white saddle on the caudal peduncle. JP: Red to brown with two white stripes and a black midbody blotch. Blotch is sometimes absent or faded (bottom right). Maximum size c.30 cm.

Abundance, Distribution & Behaviour – Common. Can be seen on most reef habitats where it swims about the area stopping to scrape algae from rocks and coral.

Common Name – **Redtail Parrotfish**

Scientific Name – *Sparisoma chrysopterum*

Local Name – Parrotfish (TP), White Bufflehead/Slut(IP)

Visual ID – TP: Green to blue with reddish crescent on caudal fin (top). This crescent is sometimes obscure. Resembles TP Yellowtail Parrotfish (next) but has large blue patch behind pectoral fin. IP & JP: Change colour rapidly from white to red to mottled and can look similar to IP Redband Parrotfish (previous) although has a thick ladder pattern on the upper and lower edges of the caudal fin (bottom) which, otherwise, is almost transparent. Maximum size c.45 cm.

Abundance, Distribution & Behaviour – Occasional, although the IP is usually common. TP has been seen around the sea rocks at Meads Bay. Seems to prefer shallow areas of coral rubble or algae beds, but is also seen swimming around reefs.

Common Name – **Yellowtail Parrotfish**

Scientific Name – *Sparisoma rubripinne*

Local Name – Parrotfish

Visual ID – TP: Green to blue with yellow crescent on caudal fin. This crescent is sometimes obscure, and may resemble TP Redtail Parrotfish (previous) but lacks large blue patch behind pectoral fin. Top image shows TP in bottom right corner, with IP in top left corner. IP & JP: Mottled grey, usually with obvious yellow caudal fin (bottom). Maximum size c.45 cm.

Abundance, Distribution & Behaviour – Uncommon, although IP is usually common. TP has been seen around the sea rocks at Meads Bay. Seems to prefer shallow areas of coral rubble or algae beds, but is also seen swimming around reefs.

Common Name – **Greenblotch Parrotfish**

Scientific Name – *Sparisoma atomarium*

Local Name – Slippery Parrot

Visual ID – TP: Shades of green to blue, mottled with faint pale to dusky or dark green blotch above pectoral fin, although there are variations (pictured). IP: Generally red but with considerable variation. JP: has a yellow wash over head and white stripes on the lower half of body. Faint yellow colourations may also be visible on base of pectoral fin. Maximum size c.12 cm.

Abundance, Distribution & Behaviour – Occasional to uncommon depending on location, seeming to prefer deeper areas, often along steeply sloping drop-offs.

Common Name – **Bucktooth Parrotfish**

Scientific Name – *Sparisoma radians*

Local Name – None

Visual ID – TP: Greenish with blue markings across eye. Black margin on caudal, anal and base of pectoral fin (top). Colourations are quite variable. IP & JP: Highly variable, changing colour rapidly to blend with background. Sometimes bluish base of pectoral fin is visible. Can be striped, bicoloured or heavily mottled (middle & bottom). Maximum size c.15 cm.

Abundance, Distribution & Behaviour – Common on seagrass beds although TP is uncommon. Younger phases are highly cryptic so are often overlooked as they hide amongst Turtle Grass blades (page 15) remaining motionless until disturbed. Otherwise seen swimming around just above the plants.

Common Name – **Bluelip Parrotfish**

Scientific Name – *Cryptotomus roseus*

Local Name – None

Visual ID – TP: Shades of green to blue with red to pink markings often forming a stripe from eye to tail (pictured). IP & JP: Rose to brown with white stripe from eye to tail. The slender body of this parrotfish helps with identification. Maximum size c.13 cm.

Abundance, Distribution & Behaviour – Occasional on seagrass beds although TP is uncommon. IP is often seen swimming around in small groups.

Family – Labridae: Wrasses & Razorfishes

Wrasses are probably the most prolific reef inhabitants with one member of the Family, the Bluehead Wrasse, considered the most common fish in the tropical western Atlantic. They are closely related to parrotfishes, but have more elongated bodies and are generally considerably smaller in size. Having said this, sizes within the Family vary; the Hogfish and the Puddingwife are able to grow to over half a metre. Some species not present in the Caribbean can achieve even greater lengths with the Humphead Wrasse (*Cheilinus undulates*) able to reach almost two and a half metres. Wrasses also differ from most parrotfishes as they do not have beaks, instead having a set of noticeable front teeth that give them a 'bucktoothed' profile. These teeth are used to break open invertebrates.

Wrasses always appear to be very busy, swimming around in loose, often mixed groups. They usually form such aggregations, sometimes consisting of hundreds of individuals, around coral heads and other raised reef structures. The Creole Wrasse is the only Family member in the Caribbean that schools in open-water, often moving between different reef habitats in search of plankton off which they feed. Some wrasses inhabit seagrass beds, and at night several species bury themselves in sand.

Like parrotfishes, as they grow up many species of wrasse first develop and mature as a female, living in a harem with a dominant male. Being sequential protogynous hermaphrodites they have the ability to change from female to male as maturity is reached. As a general rule the harems largest female will develop into a dominant male once the existing lead male is no longer present. These developmental 'phases' are split into juvenile phase (JP), initial phase (IP) and terminal phase (TP). Terminal phase are sexually mature males; initial phase includes sexually mature females and, in some species, immature and mature males; juvenile phase, as its name suggests, contains juvenile individuals. More so than with parrotfishes, although the main phases are distinct, there are usually intermediate stages where colourations blend gradually over time. The transient nature of these colour changes can make identification quite a challenge. Confounding the issue even more is the fact that there is sometimes regional variation in colouration patterns. Although many wrasses are sequential protogynous hermaphrodites as described above, others simply mature, never changing sex.

Hogfishes are part of the wrasse Family but their unique shape has rewarded them with a different common name. They use their long snout to root for food in a similar way to their namesakes. Generally hogfishes grow larger than other wrasses and their distinct colourations make them easy to identify.

Razorfishes also belong to the wrasse Family but, as with the hogfishes, their unique shape has earned them a different common name. They resemble razors in that their head is shaped rather like an upturned scalpel and their body highly compressed – both features which allow them, when frightened, to dive headfirst into the sand and tunnel away using rapid body vibrations.

Common Name – **Hogfish**

Scientific Name – *Lachnolaimus maximus*

Local Name – Hogfish

Visual ID – White with black forehead (top) but can darken to show a mottled or banded reddish brown colouration (bottom left & right); usually when foraging for food. Their dorsal fin has three elongated spines and a black blotch near the end. Younger individuals are normally paler, but distinct phases are not obvious with this species. Maximum size c.85 cm.

Abundance, Distribution & Behaviour – Uncommon. Usually seen on open sandy bottoms were it feeds so is sometimes sighted around wrecks.

Common Name – **Spanish Hogfish**

Scientific Name – *Bodianus rufus*

Local Name – Dogteeth Snapper, Piper

Visual ID – Purple upper body with yellow-gold belly and caudal fin (top). Large adults become mottled purplish-yellow. Distinct phases are not obvious with this species, although juveniles sometimes have reddish upper body (below right). Adults are also occasionally blue all over body (bottom left). Maximum size c.60 cm.

Abundance, Distribution & Behaviour – Common to occasional. Can be seen on most reef habitats but seem to favour some areas over others, especially wrecks. Juveniles look very similar to Fairy Basslets (page 153) but, behaviourally, they are very different, swimming around picking parasites and debris from larger fish.

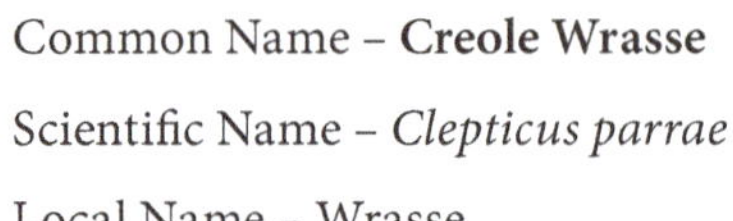

Common Name – **Creole Wrasse**

Scientific Name – *Clepticus parrae*

Local Name – Wrasse

Visual ID – TP: Dark purple body with black snout. Older individuals develop yellow to reddish areas over lower rear-body and have more hogfish-like mouth (top). IP: Similar to younger TP although is usually a lighter purple colour (bottom). JP: Purple-blue and similar to a Blue Chromis (page 143) but is easily distinguished by a series of chequered spots made up of light and dark dots running along upper back (not pictured). Maximum size c.30 cm.

Abundance, Distribution & Behaviour – Occasional but can be locally abundant in areas near deeper open-water where they swim around in large schools eating plankton. These long, streaming schools can be seen at locations such as Scrub Island, Sandy Island and Shoal Bay East; more commonly in late afternoon. Juveniles will often form groups living a more stationary existence on reefs until they reach a certain age.

Common Name – **Puddingwife**

Scientific Name – *Halichoeres radiatus*

Local Name – Slippery Parrot

Visual ID – TP: Large green to blue body with white, smudgy half-bar across mid-section (not pictured). IP: Green to blue body with four to five white vertical smudges which fade towards the belly (top). JP: Bright yellow with vivid white mid-body stripe bisected by approximately five white bars. Has black ocellated spot in middle of dorsal fin (bottom). There is an intermediate stage between the JP & IP where the ocellated spot becomes a black smudge and the vivid white middle stripe has disappeared, with bars becoming smudged and not crossing the width of the body. Maximum size c.45 cm.

Abundance, Distribution & Behaviour – Common to occasional on most reef habitats, but adults, especially TP, are uncommon to rare. Probably most abundant around the patch reefs close to Shoal Bay East point.

Common Name – **Bluehead Wrasse**

Scientific Name – *Thalassoma bifasciatum*

Local Name – None

Visual ID – TP: Blue head with two dark bars behind, separated by white area. Rear of body green to blue. IP: Often has five or six white body bars and darker overall colour (bottom). JP: Bright yellow with white belly (top). Both IP & JP have black dot on frontal dorsal fin which is important to distinguish them from Rainbow Wrasse (page 165) with which they sometimes form shoals. Body usually yellow with white belly but, occasionally, may be white or bluish. JP sometimes has dark mid-body stripe. Maximum size c.15 cm.

Abundance, Distribution & Behaviour – Abundant on most reef habitats where it swims around constantly often in large schools and may mix with other wrasse species, especially JP. JP may also mix with other species of wrasse when foraging in small groups on the bottom. TP often more solitary and swims around quickly patrolling the area. Hailed as the most abundant reef fish species in the tropical western Atlantic.

Common Name – **Yellowhead Wrasse**

Scientific Name – *Halichoeres garnoti*

Local Name – Slippery Parrots

Visual ID – TP: Yellow head and forehead with dark mid-body bar, which continues under dorsal fin, and blue to green rear-body and tail (top). IP: Dark to dusky, often bluish back with a yellow or blue belly (bottom right). JP: Yellow with brilliant blue metallic mid-body stripe (bottom left). There are a number of intermediate phase variations of the above patterns (bottom middle) but all can be recognised by eye markings varying from small dots in the juvenile to more obvious wavy lines in TP. Maximum size c.20 cm.

Abundance, Distribution & Behaviour – Common on all reefs where they swim around busily. Juveniles are also sometimes seen on or near seagrass beds.

Photo Credit Steve Donahue

Photo Credit Stuart Wynne

Photo Credit Gert-Jan Vandenbos

Photo Credit Gert-Jan Vandenbos

Photo Credit Stuart Wynne

Photo Credit Gert-Jan Vandenbos

Common Name – **Slippery Dick**
Scientific Name – *Halichoeres bivittatus*
Local Name – Slippery Parrot

Visual ID – TP: Shades of green with other rather varied colourations (top). Usually has dark stripe down mid-body reminiscent of a slippery oil streak. May have a second less distinct stripe below this. IP: Colours and markings vary greatly although usually has same dark mid-body stripe as TP (bottom left). JP: White with a distinctive dark mid-body stripe as in most other phase variations (bottom right). ALL PHASES: Green and yellow spot above the pectoral fin. This feature distinguishes it from the Blackear Wrasse (page 165). Maximum size c.20 cm.

Abundance, Distribution & Behaviour – Common on suitable habitat, although distribution varies. The TP & IP are usually occasional to uncommon swimming busily around reef and rubble areas, whereas JP favours sand, rubble or seagrass beds where it can be common to abundant. Juveniles are sometimes found in small groups with other wrasse species rummaging for food. Later phases, although preferring reef habitats, are still associated with those interspersed by sand patches.

Common Name – **Clown Wrasse**
Scientific Name – *Halichoeres maculipinna*
Local Name – Slippery Parrot

Visual ID –TP: Blue to green body with yellow to gold mid-body stripe. Has red markings on forehead, dark mid-body blotch, and another blotch on fore-dorsal fin (not pictured). IP: Pale greenish body with prominent red markings on forehead and two dark spots on the fore-dorsal fin (bottom left). An intermediate TP-IP has more developed body colour, with the dark mid-body blotch becoming apparent (top). The fore-dorsal spots begin to merge into one. JP: Dark upper side with white belly and a thin gold stripe running from snout, above eye, to caudal fin (bottom right). An intermediate IP-JP has vivid red markings on forehead, with yellow upper body stripe having less distinct upper edge that fades towards dorsal fin. ALL PHASES: Red nose (Late JP to TP: Red facial markings); these markings are, no doubt, where its common name originates. Maximum size c.15 cm.

Abundance, Distribution & Behaviour – Common on most reef habitats with JP, and sometimes IP, forming small groups with other wrasse species that travel the reef and forage for food. TP is uncommon but may be seen constantly swimming around the reef.

Common Name – **Blackear Wrasse**

Scientific Name – *Halichoeres poeyi*

Local Name – Slippery Parrot

Visual ID – TP shades of green, often with orange patches or bands (top). IP & JP vary but most individuals are bright yellow to lime green (bottom). ALL PHASES: Small dark area behind eye and tiny black spot at base of dorsal fin. These features distinguish it from the Slippery Dick (page 164). Maximum size c.20 cm.

Abundance, Distribution & Behaviour – Common to occasional, inhabiting seagrass beds and shallow reefs. Seems to prefer low profile areas and those with a more floral feel to them. Actively forages around the reef, although juveniles can be somewhat cryptic, for example, hiding amongst seagrass blades where their colour makes them very hard to see.

Common Name – **Rainbow Wrasse**

Scientific Name – *Halichoeres pictus*

Local Name – Slippery Parrot

Visual ID – TP: Various shades of blue to green on upper body with bluish mid-body stripe, pale bluish underside, and orange to red dorsal fin (bottom left). Has black blotch at base of caudal fin and blue stripes above and below eye. IP: Yellow body with white belly and dark brown mid-body stripe (bottom right). This phase is very similar to the juvenile Bluehead Wrasse (page 163), but lacks a spot on fore-dorsal fin. Intermediate TP-IP begins to develop blue facial stripes (top) and loses dark brown mid-body stripe. JP: Dark head fades down through orange to yellow towards caudal fin. Has white markings above and below the eye that again fade down through the body (not pictured). Maximum size c.15 cm.

Abundance, Distribution & Behaviour – Abundant to occasional. TP uncommon, but IP and JP can be abundant and seen in large shoals with Bluehead Wrasse when they gather above large coral heads.

Common Name – **Green Razorfish**

Scientific Name – *Xyrichtys splendens*

Local Name – None

Visual ID – TP: Varies greatly but most commonly is mottled green with one or two dark spots on mid-body (pictured). IP: Usually pale green but with no distinctive markings. JP: Highly cryptic with two elongated front dorsal spines. Some spines on the dorsal and anal fins are connected with pigmented tissues, others stand alone. ALL PHASES: Rounded caudal fin and bright red iris. Maximum size c.15 cm.

Abundance, Distribution & Behaviour – Occasional on seagrass beds and other sandy areas, often near rocks, gorgonians, or other protective cover. Usually hides in seagrass or algae, and will quickly dive into sand if threatened; this behaviour means accurate abundances are difficult to estimate. Sightings have been made on seagrass beds near Scilly Cay.

Common Name – **Rosy Razorfish**

Scientific Name – *Xyrichtys martinicensis*

Local Name – Razorfish

Visual ID – TP: Pastel greenish-blue body with pastel greenish-yellow head and dark area at base of pectoral fin (pictured). IP & JP: Pearly white head with grey to rosy body and white belly with some reddish line markings. ALL PHASES: Straight-edged caudal fin and pale iris. Maximum size c.15 cm.

Abundance, Distribution & Behaviour – Occasional, hovering above sandy areas and seagrass patches, reportedly favouring Manatee Grass (page 15). It will quickly dive into sand if threatened; this behaviour means accurate abundances are difficult to estimate.

How can I see a razorfish? Although razorfishes are notoriously hard to see while diving and snorkelling, it can sometimes pay to have a little patience. By remaining extremely still and laying on the sea bed, a persistent observer will be rewarded with fish seemingly appearing after a short time out of nowhere. It is almost impossible to make a close approach: once you have been spotted the fish will disappear in a flash, diving back under the sand. One such location where this is possible, is close to Scilly Cay in Island Harbour. But watch out for boat traffic and always make sure you have surface cover when snorkelling in port areas!

Large Eyes Reddish Bodies

Family – Holocentridae: Squirrelfishes

Squirrelfishes are so named because of their big eyes and long, pronounced dorsal fin which sticks up in a similar fashion to that of a squirrel's tail. Although most species are more active at night when they feed on invertebrates near the bottom, they are not strictly nocturnal and can be easily observed during the day, especially early morning and evening. Juvenile squirrelfish are thin silvery pelagics and seldom observed. Interestingly, many species of squirrelfishes are major sound producers, creating a resonating grunt from their air bladders, a behaviour that peaks during dusk and dawn; it is also utilised when challenging other males. Most Family members have thin, often pale body stripes.

Common Name – **Squirrelfish**

Scientific Name – *Holocentrus adscensionis*

Local Name – Redmon

Visual ID – Reddish body with light silvery stripes and occasional blotches. Fore-dorsal fin yellowish. The largest squirrelfish, reaching a maximum size of c.30 cm.

Abundance, Distribution & Behaviour – Common to occasional. Found lurking in reef recesses or shady areas during the day. Most common on shallower reefs where they often gather in groups but can be seen in deeper areas too.

Common Name – **Longspine Squirrelfish**

Scientific Name – *Holocentrus rufus*

Local Name – Redmon

Visual ID – Reddish body with silvery stripes and occasional blotches. Has row of triangular markings at tips of dorsal fin spines and the longest rear dorsal fin of all squirrelfishes. Maximum size c.30 cm.

Abundance, Distribution & Behaviour – Common on most reef habitats where it is found lurking in reef recesses or shady areas during the day. This is the most common species of squirrelfish found in Anguillian waters.

Common Name – **Reef Squirrelfish**

Scientific Name – *Sargocentron coruscum*

Local Name – Redmon

Visual ID - Reddish body with obvious silvery stripes. Has distinctive dark area on fore-dorsal fin. Tips and base of dorsal fin are white. Maximum size c.10 cm.

Abundance, Distribution & Behaviour – Uncommon. Juveniles and small adults have been sighted hiding under rocks around Sandy Island, but this species is generally wary and difficult to observe.

Common Name – **Dusky Squirrelfish**

Scientific Name – *Sargocentron vexillarium*

Local Name – Redmon

Visual ID – Dusky-grey and bronzy-red alternating stripes. Has red borders on anal and caudal fins. Maximum size c.15 cm.

Abundance, Distribution & Behaviour – Occasional. Can be locally common in some areas but their wariness makes them hard to observe. Often seen in reef areas around Sandy Island where they hide in protective recesses.

Common Name – **Longjaw Squirrelfish**

Scientific Name – *Neoniphon marianus*

Local Name – Redmon

Visual ID – Red body with orange stripes and silvery area on lower head and belly. Dorsal fin has two rows of white markings and anal fin has unusually long frontal spine. Maximum size c.15 cm.

Abundance, Distribution & Behaviour – Occasional, although can be locally common to abundant. Seen in great numbers hovering over large Finger Coral colonies around Sandy Island where this, otherwise wary, species appears unafraid. More often seen hiding in recesses during the day and seems to prefer deeper areas.

Common Name – **Blackbar Soldierfish**

Scientific Name – *Myripristis jacobus*

Local Name – Redmon

Visual ID – Vivid red body with black bar behind the head. Some white markings on fins, especially borders. Maximum size c.20 cm.

Abundance, Distribution & Behaviour – Common, although can be locally abundant in some areas, forming large schools. Often hides in reef recesses. Somewhat curious, and can usually be closely approached once their hideout has been found.

Family – Priacanthidae: Bigeyes

Bigeyes are normally found below safe-diving limits although a few species do occur in shallower waters. Of these, only the Glasseye Snapper is common in Anguilla but, due to its reclusive nature, is difficult to observe. Like other members of this Family it is nocturnal, hiding under ledges and in caves during the day. They are all thin reddish carnivores with large eyes, a continuous dorsal fin, and can grow up to 30 cm in length.

Common Name – **Glasseye Snapper**

Scientific Name – *Heteropriacanthus cruentatus*

Local Name – Bleareye

Visual ID – Red to pale silvery pink mottled body with silver bars on back, although these may be faint. Maximum size c.30 cm.

Abundance, Distribution & Behaviour – Common but their reclusive nature means they are often hard to find. Nocturnal, hiding under ledges and in reef recesses during the day. Unconcerned; once located it can be easily approached.

Is the Glasseye Snapper really solitary? These depressed-looking fish add to their sad persona by hanging out on their own under rocky recesses or overhangs for most of the day. At night they become more active when they swim around looking for food. Despite this, it is sometimes possible to see them swimming around in pairs or small groups during the day, when they pale in colour and seem to decide that it might be better to have friends after all. One place in Anguilla where this behaviour can be seen is around Anguillita, where on numerous occasions the Glasseye Snapper was observed to be a little less solitary than otherwise thought. The species, which is not a snapper even though its common name suggests otherwise, is poorly studied and little is known about its sporadic gregarious tendencies, although it is possible that is has something to do with reproduction and suggests that, at least from time to time, it does crave the company of others.

Family – Apogonidae: Cardinalfishes

Cardinalfishes are so named for their reddish colour. They are quite small with large eyes, short snouts and two separate dorsal fins. Although this Family is easy to distinguish from others, individual species are very hard to observe because, like the bigeyes, they are predominantly nocturnal and hide away in reef recesses during the day. Large groups may be seen in the caves at Scrub Island and a few species live in association with sponges, anemones and even Queen Conch (page 85), moving away from their protective lairs at night to feed in the open on tiny fish and crustaceans. A notable example of these tiny species is the **Conchfish** (*Astrapogon stellatus*), which lives in the mantle cavity of live Queen Conch (it is a mottled silver-grey colour, with numerous black speckles but no real distinguishing markings). Within the Cardinalfish Family, most species look remarkably similar but their behaviour and markings usually allow easy identification. Listed below are the species that have been confirmed to date in Anguillian waters, although regular night dives would probably uncover a number of others. Interestingly, like the jawfishes (page 176), the males of many species use their mouths to incubate the female's eggs.

Common Name – **Belted Cardinalfish**

Scientific Name – *Apogon townsendi*

Local Name – None

Visual ID – Orangey-red with dark bar between dorsal and anal fin. Has two thinner dark bars at base of tail that are sometimes fused together. Maximum size c.5 cm.

Abundance, Distribution & Behaviour – Common, although only occasionally seen during the day as it hides deep in reef recesses. May be seen peering out from entrances but is more commonly seen during night dives when they feed in the open on the reef.

Common Name – **Barred Cardinalfish**

Scientific Name – *Apogon binotatus*

Local Name – None

Visual ID – Orangey-red to salmon but can pale dramatically. Has dark bar between dorsal and anal fin, with another on caudal peduncle. Maximum size c.10 cm.

Abundance, Distribution & Behaviour – Uncommon, hiding by day in dark recesses, although occasionally seen during night dives when they feed in the open on the reef.

Common Name – **Roughlip Cardinalfish**

Scientific Name – *Apogon robinsi*

Local Name – None

Visual ID – Red body, sometimes with yellow highlights. Has faint dusky bar extending from below rear dorsal fin to belly and very faint patch across base of tail. Maximum size c.10 cm.

Abundance, Distribution & Behaviour – Uncommon, hiding by day in dark recesses, although occasionally seen during night dives when they feed in the open on the reef.

Common Name – **Sawcheek Cardinalfish**

Scientific Name – *Apogon quadrisquamatus*

Local Name – None

Visual ID – Red to bronze in colour with a dusky band at base of tail and small dark triangle under dusky eye. Maximum size c.5 cm.

Abundance, Distribution & Behaviour – Uncommon. Usually associated with anemones, as in the pictured specimens who were found living near a Corkscrew Anemone (*Bartholomea annulata* – page 46) close to the wreck of the M.V. Catheley H. Float around motionless and can be closely approached.

Common Name – **Flamefish**

Scientific Name – *Apogon maculatus*

Local Name – None

Visual ID – Bright red to salmon body with two white lines across iris. Has dark spot behind the eye, one under rear dorsal fin and another paler dark area on base of caudal fin, although this latter feature is sometimes absent. Maximum size c.10 cm.

Abundance, Distribution & Behaviour – Common, although only occasionally seen during the day as it hides in reef recesses. May be seen peering out from entrances but is more commonly seen during night dives when they feed in the open on the reef.

Small Bottom-Dwellers

Family – Gobiidae: Gobies

Gobies are small, elongate bottom-dwellers often confused with blennies. They differ somewhat in habit however, and can also be distinguished by having two separate dorsal fins. A few species drift in open-water near the reef, while others drift over their burrows in the sand or spend most of their time resting on their pectoral fins. There are probably far more members of the goby community present in Anguillian waters than are listed here but their small size, highly cryptic nature and similarities between species, mean sightings are rare and positive identification often problematic. Below are those confirmed at the time of writing.

Common Name – **Cleaning Goby**

Scientific Name – *Elacatinus genie*

Local Name – None

Visual ID – Dark upper body with pale underside. Has brilliant yellow 'V' on snout fading into pale body stripe and an underslung mouth reminiscent of a shark. Maximum size c.4 cm.

Abundance, Distribution & Behaviour – Common on all reef habitats, although its small size means it is often overlooked. Normally rests on rocks or coral, often congregating at cleaning stations. Here they perch in groups and wait for larger fish to arrive that require the removal of small parasites and other matter. There are a number of similar species throughout the Caribbean, although *E. genie* is currently the only one confirmed in Anguillian waters. Formerly classified as *Gobiosoma genie*.

Common Name – **Goldspot Goby**

Scientific Name – *Gnatholepis thompsoni*

Local Name – None

Visual ID – White body with gold spot outlined in black above pectoral fin and dark bar running from top of eye down to edge of mouth. Has scattering of dusky to dark spots over body. Fins transparent. Maximum size c.7 cm.

Abundance, Distribution & Behaviour – Occasional to uncommon, preferring sandy areas with scattered coral rubble. Usually overlooked as it blends well with surroundings while perching motionless on the bottom.

Common Name – **Pallid Goby**

Scientific Name – *Coryphopterus eidolon*

Local Name – None

Visual ID – Translucent body with yellow stripe extending from behind eye. Has faint dusky bar on base of tail and goldish blotches or stripes on body. Maximum size c.5 cm.

Abundance, Distribution & Behaviour – Uncommon, although is often overlooked because it blends so well with its background as it perches on the bottom near coral heads.

Common Name – **Bridled Goby**

Scientific Name - *Coryphopterus glaucofraenum*

Local Name – None

Visual ID – Translucent body with white 'bridle' which runs from behind mouth to the edge of its gill cover. Usually has whitish to dark spots on body and faint dusky bar on caudal peduncle. Maximum size c.7 cm.

Abundance, Distribution & Behaviour – Common, although is often overlooked because it blends so well with its background. Usually rests on small sand patches among reef areas. It is often so common that all sand patches will have at least one individual present, being almost completely invisible until darting off when closely approached.

Common Name – **Masked/Glass Goby**

Scientific Name – *Coryphopterus personatus/hyalinus*

Local Name – None

Visual ID – Bright orange to translucent body with dusky snout and a series of small white rectangles along lateral line. There is no reliable way to distinguish these two species underwater; close examination is required. Maximum size c.3 cm.

Abundance, Distribution & Behaviour – Common, although can be locally abundant when they hover in small to large groups near sheltered areas of reef. Generally, the Masked Goby is thought to inhabit water shallower than 10 m, whereas the Glass Goby is more common in water deeper than 10 m.

Common Name – **Peppermint Goby**

Scientific Name – *Coryphopterus lipernes*

Local Name – None

Visual ID – Yellow translucent body with metallic blue wash over snout. Pale stripes are often present behind eyes and second dorsal spine is considerably longer than the others. Maximum size c.3 cm.

Abundance, Distribution & Behaviour – Occasional. Is often overlooked because of its small size and ability to blend so well with its background as it perches on coral heads.

Families – Labrisomidae, Chaenopsidae & Blenniidae: Blennies

Blennies are small elongate bottom-dwellers often confused with the gobies. They differ somewhat in habit however, and can also be distinguished by having either one single or three separate dorsal fins. There are four Families of blennies, although only three are mentioned in this book, because the triplefins (Tripterygiidae) are tiny, uncommon individuals and are extremely difficult to spot. The other three Families (all with a single dorsal fin) are: scaly blennies (Labrisomidae), which include the Hairy Blenny and the Saddled Blenny; tube blennies (Chaenopsidae), which include the tiny Spinyhead, Secretary and Roughhead Blennies; and combtooth blennies (Blenniidae), which includes the Redlip Blenny.

Most blennies have fleshy appendages above their eyes called cirri, sometimes with additional structures on their snout. Blennies have the ability to change colour and pattern rapidly, enabling them to blend remarkably well with their surroundings, making them a challenge to find. Inter-species similarities, combined with these colour changes, often makes identification problematic; this is further complicated by the fact that sometimes males and females of the same species can be dramatically different. These factors, combined with their small size, mean that there are probably far more members of the blenny community present in Anguillian waters than listed here. Below are those confirmed at the time of writing.

Common Name – **Hairy Blenny**

Scientific Name – *Labrisomus nuchipinnis*

Local Name – Sorearse

Visual ID – Brownish black body with dark spot on upper gill cover. Has thick lips and is the largest blenny found anywhere in the Caribbean, growing up to a maximum of c.20 cm.

Abundance, Distribution & Behaviour – Uncommon, preferring a wide range of shallow habitats from rocky areas to gravel, or even seagrass beds. Has been seen in the shallow reef close to Katouche Bay and around the sea rocks in Maundays Bay.

Common Name – **Saddled Blenny**

Scientific Name – *Malacoctenus triangulatus*

Local Name – None

Visual ID – White to pale body with four dark inverted triangular saddles across back. Has dark bar on caudal peduncle. Colourations vary. Maximum size c.5 cm.

Abundance, Distribution & Behaviour – Common, although its small size and cryptic nature mean it is easily overlooked. Inhabits most reef areas, and can also be found hiding between seagrass blades.

Common Name – **Spinyhead Blenny**

Scientific Name – *Acanthemblemaria spinosa*

Local Name – None

Visual ID – Dark body with white snout covered in short cirri. Yellow iris. Similar species include the **Secretary Blenny** (*Acanthemblemaria maria* – not pictured) which has paler body and lacks white snout. Maximum size c.3 cm.

Abundance, Distribution & Behaviour – Common to occasional. Often overlooked because of their small size and habit of remaining in burrows with only their head protruding. Their cirri act as sense organs, and when small plankton brushes past they extend out of their burrows and consume it.

Common Name – **Redlip Blenny**

Scientific Name – *Ophioblennius macclurei*
(formerly *O. atlanticus*)

Local Name – None

Visual ID – Dark reddish brown with blunt head and large red lips. Maximum size c.12 cm.

Abundance, Distribution & Behaviour – Common in some areas, preferring rocky inshore habitats and shallow reefs. They rest on their pectoral fins and are often seen in pairs. Can be locally abundant and are probably the most easily observable species of blenny Anguillian waters.

Family – Opistognathidae: Jawfishes

Jawfishes, as their name suggests, have large mouths and big jaws. They use these jaws to move stones and sand creating burrows, leaving only their heads protruding once inside. Their large mouth also comes in handy when the male uses it to incubate the female's eggs, a process that takes about a week and leaves the father unable to eat until their birth. It shares this behaviour with some species of cardinalfishes (page 170-171). When the fish come out from the protection of their burrows, they hover close by while feeding on plankton and other small organisms. At the first sign of danger they dart back into their burrows, with just their head once again protruding. If closely approached even by the most stealthy diver, they disappear completely until they feel it is safe enough to rise back up and peer out from the burrow entrance to see if the coast is clear.

Common Name – **Yellowhead Jawfish**

Scientific Name – *Opistognathus aurifrons*

Local Name – None

Visual ID – Pale slender body with yellowish head. Identification is facilitated by behaviour. Maximum size c.10 cm.

Abundance, Distribution & Behaviour – Occasional on sand and coral rubble near reefs. Individuals hover vertically above their burrows, into which they retreat tail-first if frightened.

Common Name – **Banded Jawfish**

Scientific Name – *Opistognathus macrognathus*

Local Name – None

Visual ID – Light brown head with mottled, usually indistinct, markings. When in its burrow it is difficult to distinguish from the Dusky Jawfish (*Opistognathus whitehursti* – not listed in this book). The Banded Jawfish lines the entrance of its burrow with shells and stones, whereas the Dusky Jawfish does not. Maximum size c.20 cm.

Abundance, Distribution & Behaviour – Occasional on sand, seagrass beds and coral rubble near reefs. Usually sits in its burrow with only its head protruding, although will retreat inside if threatened. A number of individuals have been seen in Little Bay and Meads Bay.

Family – Callionymidae: Dragonets

The Dragonets are a diverse Family worldwide, although only a small number of species live within the Caribbean. Of these, only one species has been confirmed in Anguillian waters. Sightings are infrequent because it is able to change colour; lying motionless on the bottom blending perfectly with its background. Due to similarities in morphology and behaviour, dragonets are sometimes confused with members of the goby Family.

Common Name – **Lancer Dragonet**

Scientific Name – *Paradiplogrammus bairdi*

Local Name – None

Visual ID – Pale brownish mottled body with elongated front dorsal fin and large goggle eyes. Often has scattering of metallic blue ocellated spots. The female's elongated dorsal fin is usually shorter and darker than that of the male. Maximum size c.5 cm.

Abundance, Distribution & Behaviour – Uncommon on reefs and close-by areas of sand and rubble, although their true abundance is difficult to assess due to their highly cryptic nature. The pictured specimen was sighted laying motionless among algae coated pebbles in the channel between Scrub Island and Little Scrub.

Odd-Shaped Bottom-Dwellers

Family – Bothidae: Lefteye Flounders

Lefteye flounders derive their name from the fact that they lie on the sea floor on what used to be their right side; over evolutionary time the eye on their right side has migrated round to the left side, thus both eyes are now on the left. Interestingly the larvae look and develop much like fish with regular body plans until a certain age is reached when the right hand eye begins to move around to the other side. In this strange altered body arrangement, designed probably to avoid predators and also aid in prey capture, the pectoral fins have not altered position, and the dorsal fin and anal fins have elongated to form what now look like marginal body frills. Of the species found in the Caribbean, only one has been confirmed in Anguillian waters.

Common Name – **Peacock Flounder**

Scientific Name – *Bothus lunatus*

Local Name – Platefish

Visual ID – Lightish brown, although can pale dramatically to blend with most backgrounds. Has numerous blue rings all over body with spots over fins and head. These blue rings and spots can also lighten and darken. Maximum size c.45 cm.

Abundance, Distribution & Behaviour – Occasional, although their habit of lying camouflaged while half-buried in sand means true abundance is difficult to assess. Prefers sand and seagrass areas, but are also found on reef habitats where they are more easily visible. Peacock Flounder spend most of their time on the bottom blending with their background, only moving when closely approached.

Family – Scorpaenidae: Scorpionfishes

Scorpionfishes are so called because of toxic spines on their back and fins which are connected to venom sacs. They have large stout heads, and their tassels, warts and ability to change colour mean that they are masters of disguise and so usually overlooked. Living on the bottom, in crevices and under overhangs they feed on crustaceans, fishes and other unsuspecting passers-by. Luckily for divers and snorkellers scorpionfishes are not aggressive, although they will erect their spines if threatened, a habit that often gives away their position and avoids a nasty encounter with unsuspecting feet or hands. Although the species found around Anguilla are not known to be fatally poisonous, a sting can cause severe pain and illness. In August 2010, the Red Lionfish (opposite), a new member of this Family arrived in Anguilla, following its introduction into the Caribbean in Florida. With no natural predator this species rapidly colonised the area and is now the target of eradication efforts by dive operators and the Department of Fisheries and Marine Resources. The species is also tasty to eat, so the efforts of local fishers also help to keep numbers down. It is hoped that if numbers remain low for long enough, local species (for example Groupers) may learn techniques to consume this toxic foe, as they do in their native Indo-Pacific region. Two species were reportedly introduced in Florida, *Pterois miles* and *Pterois volitans*, although only the latter has been confirmed in Anguillian waters. Currently populations of this species are closely monitored, so please report any sightings directly to The Department of Fisheries and Marine Resources.

Common Name – **Spotted Scorpionfish**

Scientific Name – *Scorpaena plumieri*

Local Name – Lionfish

Visual ID – Highly mottled body patterns usually a mixture of brown, red and tan. Numerous fleshy head and body flaps, and sometimes large cirri above eyes and on chin. Caudal fin has three dark bars. Upper side of pectoral fins (only visible when swimming) has dark area with brilliant white spots and purple to red rays. Maximum size c.45 cm.

Abundance, Distribution & Behaviour – Once the most common scorpionfish on Anguilla's reef areas, this species was often overlooked due to its habit of sitting motionless on the bottom, blending with the background. Although favouring reefs, it can be found on all habitat types. Puncture wounds from venomous fore-dorsal fin spines can cause severe pain and illness.

Common Name – **Mushroom Scorpionfish**

Scientific Name – *Scorpaena inermis*

Local Name – Lionfish

Visual ID – Generally mottled red and brown with two faint bars on its caudal fin and dark margins on pectoral, rear dorsal and anal fins. Also has rather inconspicuous upside-down, mushroom-like growths extending over upper eye. This is the smallest species of scorpionfish in the Caribbean and rarely exceeds 12 cm in length.

Abundance, Distribution & Behaviour – Uncommon, although cryptic nature makes sighting less likely, often hiding under rocks or other reclusive places. Seems to favour areas of gravel or rocks rather than solid reef. This non-venomous species of scorpionfish lays motionless blending with background, only moving if molested.

Common Name – **Red Lionfish**

Scientific Name – *Pterois volitans*

Local Name – Lionfish

Visual ID – Long, elaborate venomous spines fan out from pectoral and dorsal fins, with red and white zebra striping over the body. Fleshy 'tentacles' protrude from above the eyes and below the mouth. Distinctive. Maximum size c.40 cm.

Abundance, Distribution & Behaviour – Now the most common scorpionfish found on Anguilla's reef areas, the Red Lionfish is usually seen hovering motionless over reef areas or under overhangs. Local population hotspots have been identified, including Crocus Bay, Meads Bay, Dog Island and Anguillita. Puncture wounds from venomous fin spines can cause severe pain and illness, but rarely prove fatal.

Family – Antennariidae: Frogfishes

Frogfishes, also sometimes referred to anglerfishes, are stout strange-looking fish. They have a down-turned mouth and modified dorsal spine with a flap of skin at the apex which acts as a lure to attract prey. Masters of camouflage, changing their bodies to almost any colour, the frogfish sits in wait waving its lure, and any prey attracted to it quickly gets snapped up. The shape of their mouth allows it to be opened to the width of their bodies so engulfed prey can often be quite large. Although a number of frogfish species are present in the Caribbean, only one is thought to occur in Anguillian waters – but its highly cryptic nature means sightings are very rare, so abundance is not currently known.

Common Name – **Longlure Frogfish**

Scientific Name – *Antennarius multiocellatus*

Local Name – None

Visual ID – Extremely varied in colour, with many different phases and patterns that blend almost perfectly with background. Has long translucent 'lure' just above lip, and three, sometimes ocellated, spots on its caudal fin. Very distinctive body shape. Maximum size c.20 cm.

Abundance, Distribution & Behaviour – Presumed common due to reports from around the region, but its incredibly effective cryptic nature means it has yet to be confirmed in Anguilla. It inhabits reef areas and often rests on and blends in perfectly with sponges. Will not move unless directly molested, apparently relying on its camouflage.

Family – Cirrhitidae: Hawkfishes

Hawkfishes, similar in many ways to the hamlets, are usually seen in crevices waiting to pounce on unsuspecting prey. However, unlike other ambush predators they are not highly cryptic and will often move from place to place. There is only one species of hawkfish found in the Caribbean.

Common Name – **Redspotted Hawkfish**

Scientific Name – *Amblycirrhitus pinos*

Local Name – None

Visual ID – Brownish body with white bars and caudal fin. Has red spots over head, upper body and dorsal fin. Maximum size c.10 cm.

Abundance, Distribution & Behaviour – Uncommon, although its reclusive nature and general wariness mean it may be far more abundant than thought. Prefers coral areas where it perches on the bottom, often in recesses, occasionally darting about.

Family – Dactylopteridae: Flying Gurnards

Flying gurnards spend their time walking around on the finger-like spines of their lower pectoral fin, turning over rubble to look for prey. They have a hard head and upper body (similar in durability to the body of the boxfishes) and large pectoral fins that often lead to the confusion, accentuated by their name, that flying gurnards can fly. In fact they are unable to do so and only seem to use these fins to startle predators, as when alarmed they rapidly spread them out and swim away. Quite closely related to this Family, the searobins (Triglidae) also walk around on similar finger-like spines, but have yet to be confirmed in Anguillian waters.

Common Name – **Flying Gurnard**

Scientific Name – *Dactylopterus volitans*

Local Name – Batfish

Visual ID – Grey to yellow-brown with white spots on upper body. Has blunt snout and ability to pale or darken dramatically. Has huge fan-like pectoral fins with metallic blue line and dot markings close to margin. Maximum size c.45 cm.

Abundance, Distribution & Behaviour – Uncommon, living mainly on sand patches or among coral rubble where they use their separate, finger-like lower pectoral fin rays to 'walk' around and dig for food. The main fan-like pectoral fins are usually folded against their side, only being extended if alarmed. Recent sightings have been made in Little Bay and Maundays Bay.

Family – Synodontidae: Lizardfishes

Lizardfishes have large down-turned mouths and are bottom-dwellers expert in camouflage. Some species bury themselves in sand with just their heads protruding and feed on buried invertebrates or wait on other unsuspecting prey. Although lizardfishes are similar in markings and colourations the two species listed here can be easily identified by the depths at which they are found and their behaviour.

Common Name – **Sand Diver**

Scientific Name – *Synodus intermedius*

Local Name – Lizardfish

Visual ID – Usually has pale body with reddish brown bars across back. Has yellow to gold or blue stripes with diamond pattern on side and can darken dramatically. Resembles the Inshore Lizardfish (next) but markings usually converge with each other. Maximum size c.40 cm.

Abundance, Distribution & Behaviour – Common on most reef habitats, preferring those with sandy areas where they rest on the bottom or partially bury themselves, sometimes with only the head visible. The Sand Diver is rarely seen in water shallower than 5 m, which is again a useful guide in distinguishing it from the Inshore Lizardfish.

Common Name – **Inshore Lizardfish**

Scientific Name – *Synodus foetens*

Local Name – Lizardfish

Visual ID – Pale body with dusky bars across back and diamond pattern on side. Can darken dramatically to resemble the Sand Diver (previous) but markings do not converge with each other. Maximum size c.25 cm.

Abundance, Distribution & Behaviour – Uncommon, although is frequently spotted in the sandy shallows at beaches such as Meads Bay and Rendezvous Bay, where it lies on the bottom and partially buries itself with only the head protruding.

Family – Syngnathidae: Pipefishes & Seahorses

Pipefishes and seahorses belong to a highly charismatic fish Family whose uniqueness often links them to fairytales and other fantasy stories. In terms of fish however they might be considered quite strange-looking with trumpet-like snouts and small mouths. They are encased in, usually visible, protective body rings. Seahorses are vertically orientated with a cocked, horse-like head and coiled tail which is often used to anchor onto plants and corals. Pipefish, as their name suggests, are narrow elongate individuals with a forward pointing head. All Family members have lost, or at least highly modified, the vast majority of their fins. Although a seahorse sighting was made by one of the dive operators here in Anguilla a number of years ago, the precise species has not been confirmed. The Longsnout Seahorse has been listed here as it is the most probable species to be present.

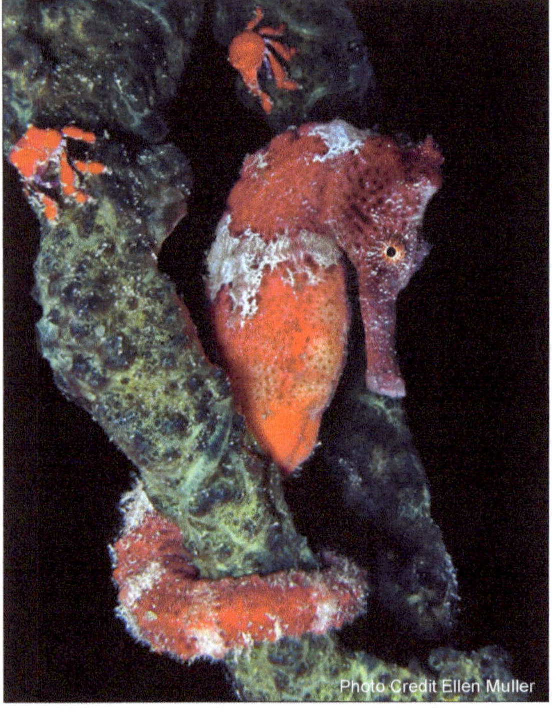

Common Name – **Longsnout Seahorse**

Scientific Name – *Hippocampus reidi*

Local Name – Seahorse

Visual ID – Body colour varies greatly but usually has small black speckles over head and body. Maximum size c.10 cm.

Abundance, Distribution & Behaviour – Unknown. Small size and often cryptic nature means they can be easily overlooked, often attaching themselves to gorgonians or seagrass (or are associated with free-floating sargassum - page 16). May be abundant in localised areas, such as under docks. Although not positively identified, a black-coloured seahorse was found by children playing under the wooden pier at Sandy Ground. Any positive sightings should be reported to the Department of Fisheries and Marine Resources.

Common Name – **Shortfin Pipefish**

Scientific Name – *Cosmocampus elucens*

Local Name – None

Visual ID – Body colour varies from dark brown to lavender, with numerous pale body bands. Maximum size c.15 cm.

Abundance, Distribution & Behaviour – Uncommon. Is most likely to be spotted swimming though seagrass blades in areas such as Little Bay. Can also be found in reef habitats although it is highly cryptic, swimming through small tunnels and reef recesses, and so rarely seen.

Odd-Shaped Swimmers

Family – Aulostomidae: Trumpetfishes

Trumpetfishes are a small Family of fish with only three species worldwide, all belonging to the same Genus. They are an elongated tube-shaped fish that can reach quite large sizes and are closely related to the seahorses (previous), sharing similar interwoven struts of bone which support the body, keeping it fairly rigid. Trumpetfishes are carnivorous and hover motionless, blending well with their background, then slowly approach their prey. When close enough they strike, rapidly expanding their trumpet-like mouth almost to the diameter of their body, creating a suction force that draws in their meal. It should be mentioned that cornetfishes, though not thought to be present in Anguillian waters, are very similar in appearance but belong to a different Family.

Common Name – **Trumpetfish**

Scientific Name – *Aulostomus maculatus*

Local Name – Trumpetfish

Visual ID – Long thin body of varying colours. Most common phase is light brown with pale lines and a scattering of black dots. Can also be blue (especially the head), bright yellow (below right) and all the colours in-between. Maximum size c.1 m.

Abundance, Distribution & Behaviour – Common on most reef habitats, although its habit of remaining still, sometimes hovering vertically mimicking sea rods (below left), or hiding inside other flora (middle right) means it can easily be missed. May also attempt camouflage by remaining very close to larger fish as they move about the reef. This behaviour presumably fools both predator and prey alike.

Family – Malacanthidae: Tilefishes

Tilefishes seek refuge in caves, self-made burrows, or under piles of rock, and feed on small crustaceans, molluscs and worms. They have a long slender body with anal and dorsal fins that run almost its entire length. Tilefishes spend most of their time hovering close to their burrows using keen eyesight to spot prey. Only one species is found in the Caribbean.

Common Name – **Sand Tilefish**

Scientific Name – *Malacanthus plumieri*

Local Name – Whitening

Visual ID – Long, pale, usually white body with crescent-shaped caudal fin that has yellow margins and a dark patch on its upper lobe. Often has light yellow and blue markings over the head. Maximum size c.60 cm.

Abundance, Distribution & Behaviour – Common on sandy rubble patches where it builds its burrow. Has a habit of hovering a little above the surface of the sand close to the entrance of its burrow, into which it will disappear if threatened.

Family – Tetraodontidae: Pufferfishes

Pufferfishes get their name from a rare ability (shared only with the porcupinefishes – next) to inflate their body as a defence mechanism by drawing water into their highly flexible stomachs. Their Family name, Tetraodontidae, relates to four fused teeth on the upper jaw which, when combined with their powerful jaws, are used to crush hard-shelled invertebrates for food. They differ from the closely related porcupinefishes because they lack spines. Instead, the skin has a rough sandpaper-like feel, similar to that of a shark. Pufferfishes swim with a combination of their pectoral, dorsal, anal and caudal fins making them highly manoeuvrable, yet slow, and is probably the reason for their primary defence mechanism. As a secondary defence, pufferfishes are toxic, and are in fact the second most poisonous vertebrate to humans in the world, although their toxicity to potential predators remains to be tested. Most species are toxic, although they are prized in a dangerous Japanese-based culinary practice which removes the toxic skin (mucus) and internal organs before consumption. The two species commonly found in Anguillian waters are small and no cases of poisoning have been reported here, although if unintentionally caught fishers are advised to wear gloves and not consume the catch.

Common Name – **Sharpnose Puffer**

Scientific Name – *Canthigaster rostrata*

Local Name – None

Visual ID – Brown upper body, yellowish-golden belly and caudal fin. Belly has blue speckles, and caudal fin a dark border. Blue line markings are present near eye. Maximum size c.10 cm.

Abundance, Distribution & Behaviour – Common on most reef habitats and also sometimes on seagrass beds. Will nibble on plant tips, and also eat a wide range of small invertebrates.

Common Name – **Bandtail Puffer**

Scientific Name – *Sphoeroides spengleri*

Local Name – None

Visual ID – Light brown upper body with speckles. Has row of dark blotches on white belly and two dark bands on caudal fin. Maximum size c.25 cm.

Abundance, Distribution & Behaviour – Occasional on seagrass beds and areas of coral rubble. Rarely seen on reef habitats. Will often hover near the bottom blending with background making sightings difficult until individual glides slowly away. Sightings quite common in the rubble areas at Limestone Bay.

Common Name – **Checkered Puffer**

Scientific Name – *Sphoeroides testudineus*

Local Name – None

Visual ID – Pale yellow to tan body with numerous relatively large defined brown patches on back and sides. Underside is white and unmarked. Maximum size c.10 cm.

Abundance, Distribution & Behaviour – Uncommon, although its preference for sitting camouflaged on the sea floor amongst seagrasses means it is often overlooked, especially if partially buried. Prefers sheltered shallow bays. Sightings to date have all been made in Little Bay.

Family – Diodontidae: Porcupinefishes

Porcupinefishes are so called because of spines that are present all over their body. They have the rare ability (shared only with the pufferfishes – previous) to inflate their body as a defence mechanism by drawing water into their highly flexible stomachs. In species where spines are not permanently erect, this inflation process erects the spines and shocks the predator into rethinking lunch plans. The porcupinefishes have large eyes and chubby lips that give them a cute, cartoon character-like appearance.

Common Name – **Bridled Burrfish**

Scientific Name – *Chilomycterus antennatus*

Local Name – Hedgehog

Visual ID – Yellow-brown body with small black spots and some larger, darker, often ocellated patches. In the featured image this patch is above the pectoral fin but quite pale. Spines are always erect. Has a golden iris with circle of black dots. Two 'antennae' are often visible, one over each eye. Maximum size c.30 cm.

Abundance, Distribution & Behaviour – Uncommon to rare, inhabiting seagrass beds and sometimes adjacent reefs. Will often hide in reef recesses or seek other cover if approached, but on rare occasions are seen peering inquisitively from the entrance. More active at night.

Common Name – **Spotted Burrfish**

Scientific Name – *Chilomycterus atringa*

Local Name – Hedgehog

Visual ID – Uneven shades of brown on body with white to pale belly. May display large dark patches. Body and all fins covered in small dots. Spines are triangular and always erect. Maximum size c.40 cm.

Abundance, Distribution & Behaviour – Uncommon to rare, usually hiding by day in shaded areas, caves and other protective recesses, but on rare occasions are seen peering inquisitively from the entrance. More active at night.

Common Name – **Balloonfish**

Scientific Name – *Diodon holocanthus*

Local Name – Hedgehog

Visual ID – Pale brown body with darkish patches and brown band between eyes. Has black spots over body and spines which lie flat, only being raised when inflating defensively. Maximum size c.50 cm.

Abundance, Distribution & Behaviour – Uncommon, inhabiting seagrass areas and reefs. Often hide in protective recesses by day but are sometimes seen peering inquisitively from the entrance. More active at night.

Common Name – **Porcupinefish**

Scientific Name – *Diodon hystrix*

Local Name – Hedgehog

Visual ID - Pale brown body with black spots over body and fins. Spines lie flat, only being raised when body is inflated. Maximum size c.1 m

Abundance, Distribution & Behaviour – Occasional on most reef habitats, although on rare occasions is seen on seagrass areas. Usually hides in protective recesses during the day but are sometimes seen peering inquisitively from the entrance. May be seen free-swimming by day although is generally more active at night and commonly seen on night dives.

Family – Ostraciidae: Boxfishes

Boxfishes, with their small protruding mouths and pouting lips, are so called because their body is, essentially, a bony box of armour fitted together in a series of hexagonal plates derived from scales. Unlike a box, however, they have a flat ventral surface with sides that slope up to a dorsal 'apex'. Like the pufferfishes they move predominantly with their pectoral and dorsal fins, which provide good manoeuvrability, but put on an impressive turn of speed with a flick of their splayed caudal fin. Most species can lighten and darken dramatically; the Honeycomb Cowfish, for example, is often dark as it browses lazily over a seagrass bed, suddenly becoming a brilliant white if startled. Juvenile boxfishes, which often vary greatly from the adults, are usually round in shape and sometimes lovingly referred to as peafish.

Common Name – **Scrawled Cowfish**

Scientific Name – *Acanthostracion quadricornis*

Local Name – Shellfish

Visual ID – Yellow to blue-green body with scrawled pattern of bluish markings over body. Has two sharp spines above each eye and two below anal fin. Maximum size c.40 cm.

Abundance, Distribution & Behaviour – Uncommon to rare, living in a wide range of habitats from seagrass to reefs. Can change colour dramatically and often looks as if it is shimmering. Appears to rely on camouflage, only retreating when it knows it has been detected. Has been seen on the seagrass bed in front of Scilly Cay and out on Prickly Pear reef.

Common Name – **Honeycomb Cowfish**

Scientific Name – *Acanthostracion polygonia*

Local Name – Shellfish

Visual ID – Bluish-Silver to yellow but can change colour dramatically becoming almost black. A hexagonal or honeycomb pattern is usually visible, at least in part, on body. Has two sharp spines above each eye and two below anal fins. Juvenile is orange with dusky spots (bottom). Maximum size c.40 cm.

Abundance, Distribution & Behaviour – Occasional on all reef habitats but may also be seen foraging on seagrass beds. Relies on camouflage and often hovers motionless until it is sure it has been detected, when it quickly darts off. May alter colour to startle predators, rapidly becoming white if aggressive movements are perceived.

Common Name – **Trunkfish**

Scientific Name – *Lactophrys trigonus*

Local Name – Sheephead, Shellfish

Visual ID – Usually pale but colour varies. Young individuals have covering of white spots. A dark-edged honeycomb pattern develops with age, and older individuals sometimes exhibit a scrawled pattern. Has two spines below anal fin and an obvious hump on back. Maximum size c.45 cm.

Abundance, Distribution & Behaviour – Uncommon, but may be locally occasional. Swims over seagrass beds and sometimes over reefs foraging for food. Have been sighted in Maundays Bay and also on the seagrass bed in front of Scilly Cay. This latter area is where the pictured specimen was landed by fishers.

Common Name – **Smooth Trunkfish**

Scientific Name – *Lactophrys triqueter*

Local Name – Shellfish

Visual ID – Dark body covered with white spots, although central body is paler with honeycomb pattern present. Is dark around mouth and the only member of the Family with no spines. Juvenile has black body with yellow spots; the honeycomb patch developing with age. Maximum size c.30 cm.

Abundance, Distribution & Behaviour – Common on most reef habitats and occasionally over sand or seagrass beds. Swims around unconcerned, foraging for food. Juvenile is the most commonly seen immature boxfish in Anguilla, usually seen swimming around the crevices of small coral heads.

Common Name – **Spotted Trunkfish**

Scientific Name – *Lactophrys bicaudalis*

Local Name – Shellfish

Visual ID – White body covered with black spots. Is white around the mouth and has two spines below anal fin. Maximum size c.40 cm.

Abundance, Distribution & Behaviour – Occasional, often living under reef ledges or swimming around the entrance of other reef recesses

Family – Balistidae: Triggerfishes

Triggerfishes are known, along with filefishes (next), as 'leatherjacks' because of their thick rough skin. They have an elongated dorsal spine, that can be raised or lowered at will, which is locked into place by a rigid second spine. The tissue that connects these spines resembles a trigger, hence their common name. Their leathery skin allows them to feed on sea urchins and other well defended creatures, and their trigger inhibits potential predators from making an easy meal of them, as it wedges into the corner of their mouths. Triggerfishes are amongst the most elegant reef dwellers, swimming along by gracefully undulating their dorsal and anal fins.

Common Name – **Queen Triggerfish**

Scientific Name – *Balistes vetula*

Local Name – Old Wife

Visual ID – Purple to pale blue body but colour can vary. Has beautiful streaming tips of rear dorsal and caudal fin, with two blue stripes on face and dark lines radiating from eye. Can change colour dramatically. Younger individuals have more dramatic colourings but less elongated fins (lower image). Maximum size c.60 cm.

Abundance, Distribution & Behaviour – Occasional to uncommon in most reef areas, although may still be common in some localities where fishing has presumably had less of an impact on their populations. They swim about the reef almost constantly looking for food and are especially keen on sea urchins.

Common Name – **Ocean Triggerfish**

Scientific Name – *Canthidermis sufflamen*

Local Name – Turbit

Visual ID – Pale in colour with a black blotch at base of pectoral fin. Can darken dramatically. Maximum size c.60 cm.

Abundance, Distribution & Behaviour – Uncommon on most reef habitats. Normally swims in loose groups but is sometimes solitary. If seen close to reefs they favour areas near open-water. On a number of occasions has been sighted at Sandy Island and near Seal Island; can often be found around wrecks.

Common Name – **Sargassum Triggerfish**

Scientific Name – *Xanthichthys ringens*

Local Name – None

Visual ID – Bluish to brownish grey with three dark lines on cheek and a series of dashes forming thin stripes on body. Can pale and darken dramatically and has a white dot in front of eye. A black border extends from base of dorsal and anal fins to caudal fin. Maximum size c.25 cm.

Abundance, Distribution & Behaviour – Rare. May be common in deeper regions, for example, Sombrero Island and unconfirmed reports from close to Dog Island. Reportedly prefers low profile sections of outer reefs.

Common Name – **Black Durgon**

Scientific Name – *Melichthys niger*

Local Name – Pigfish

Visual ID – Black body, possibly with bluish cast. Has pale lines along base of dorsal and anal fins, and blue markings around eye. Can change colour to become pale. Maximum size c.40 cm.

Abundance, Distribution & Behaviour – Locally abundant on offshore reef areas, especially the cays. Swim in loose aggregations some distance over the reef, although may hide in reef recesses when threatened, locking themselves in by raising their trigger. Are not usually present around inshore areas.

Are triggerfish aggressive? In terms of facial features, most members of the triggerfish Family look as though they could get quite cross if their space isn't respected. Indeed, although usually wary of divers, some species of triggerfish in other parts of the world have a steely reputation for seeing off divers who enter their territory, especially during breeding seasons. The Titan Triggerfish (*Balistoides viridescens*) is one such example from the Indo-Pacific region. As its common name rightly suggests, this fish is a force to be reckoned with, and females will pursue unwanted guests with great vigour. They have even been known to bite chunks from a fleeing divers fin! Luckily for underwater enthusiasts in Anguilla, Caribbean species do not normally exhibit such behaviour. There have however been a small number of documented cases where the Queen Triggerfish (page 189) has taken exception to intruders and taken decisive action to make her feelings known. As with all marine life, it is always best to be cautious and treat underwater creatures with the utmost respect.

Family – Monacanthidae: Filefishes

Filefishes are closely related to triggerfishes (previous) with both Families collectively known as the leatherjackets. Filefishes skin was once used as a type of sandpaper to finish wooden boats. They differ from the triggerfishes in that they cannot lock their dorsal 'trigger', are more elongated and laterally compressed and have a similar pouting mouth to that of the boxfishes. Their lateral compression allows them to squeeze into narrow reef crevices, and some have a habit of hanging vertically among algae and octocorals in a similar way to the trumpetfishes (page 183), presumably to fool both predator and prey alike. Certain species have a flap of skin with an articulated bone which can be moved up and down to form an appendage known as the dewlap; this is used to make the fish appear bigger but may be used simultaneously with the dorsal spine to make it more difficult for a predator to remove the fish from a crevice. Filefishes have a similar diet to that of the boxfishes, eating small hard-shelled invertebrates. Family members vary greatly in size from the large Scrawled Filefish possibly exceeding one metre in length to the tiny Slender Filefish that rarely exceeds 5 or 6 cm.

Common Name – **Scrawled Filefish**

Scientific Name – *Aluterus scriptus*

Local Name – Bastard Old Wife

Visual ID – Pale yellow to dark brown in colour but can pale or darken dramatically. Has blue to blue-green stripes or dots over the elongated body and a broom-like tail. Maximum size c.1 m.

Abundance, Distribution & Behaviour – Uncommon, although can be found in a variety of habitats; may drift slowly over reefs, in open-water, hang vertically under docks or be found on seagrass beds.

Common Name – **Orangespotted Filefish**

Scientific Name – *Cantherhines pullus*

Local Name – Bastard Old Wife

Visual ID – Brown body with white spot on base of caudal fin. Orange to yellow spots form stripes that become bolder towards tail. Can darken and lose spots altogether, but bluish line markings on head remain. Maximum size c.20 cm.

Abundance, Distribution & Behaviour – Common on most reef habitats, although they tend to be shy, staying near the bottom or hiding among octocorals.

Photo Credit Lippold Hakken

Common Name – **Whitespotted filefish**

Scientific Name – *Cantherhines macrocerus*

Local Name – Bastard Old Wife

Visual ID – Shades of grey to orange or brown with whitish snout. Can display a series of pale, whitish spots or can lack them entirely, instead displaying bold orange colourations. Maximum size c.45 cm.

Abundance, Distribution & Behaviour – Uncommon, but has been frequently seen around Dog Island. Usually swims around reef tops in pairs, each displaying a different colour pattern.

Photo Credit Stuart Wynne

Photo Credit Stuart Wynne

Common Name – **Slender Filefish**

Scientific Name – *Monacanthus tuckeri*

Local Name – None

Visual ID – Yellow to brown slender body with a white reticulated pattern. Can pale or darken obscuring these markings. The dewlap is large and usually has yellow edge with thin blue markings and dark central area. Maximum size c.8 cm.

Abundance, Distribution & Behaviour – Occasional on reef habitats, although its cryptic ability to blend with octocorals often make it difficult to spot. May also be found on seagrass beds. The dewlap is often extended which is possibly a defensive posture.

Photo Credit Stuart Wynne

Common Name – **Fringed Filefish**

Scientific Name – *Monacanthus ciliatus*

Local Name – None

Visual ID – Body colour is highly variable but usually shades of greenish-grey to brownish-yellow. Can pale or darken dramatically. Has a large dewlap and skin appears rough, almost as if algae is growing on it. Maximum size c.15 cm.

Abundance, Distribution & Behaviour – Common on seagrass beds, although its cryptic nature means it is very difficult to spot. May also be found on rubble areas, but prefers those with vegetation. Often drifts vertically within clumps of algae.

Family – Pempheridae: Sweepers

Sweepers belong to a Family of small-bodied species with an unusually, somewhat distinctive, deep belly profile. Most species favour dark recesses, caves and shipwrecks during the day where they congregate in small to large groups. They have relatively large eyes and a laterally compressed body which suits this behaviour well. There is only one species from this Family known in Anguillian waters.

Common Name – **Glassy Sweeper**

Scientific Name – *Pempheris schomburgki*

Local Name – None

Visual ID - Copper colour with dark strip at base of anal fin and deep belly profile. Maximum size 15 cm.

Abundance, Distribution & Behaviour – Occasional, usually only being spotted by an inquisitive observer investigating under ledges or in small caves where they usually congregate (see note below).

Underwater caves in Anguilla: Although not particularly common, there are a few examples of underwater caves in Anguilla that are worth taking a look at, which usually contain beautiful shoals of silvery fish or schools of species such as the Glassy Sweeper (above). Such examples exist within the sea rocks at the western end of Meads Bay and Barnes Bay; through a 'secret' underwater entrance at Blowing Rock; within the small cliffs at Black Garden and Hole-in-the-Wall dive site; and in various locations around Scrub and Little Scrub Islands. Ask local dive operators for details.

Snorkelling the cave at Meads Bay

Family – Mullidae: Goatfishes

Goatfishes are so called because their barbels, that are found under the chin, resemble a goats 'beard'. They use these appendages to dig around in the sand or rubble looking for food; thus they are often accompanied by other species, such as Bar Jacks (page 110) or Plumas (page 220), who are hoping to snatch a free meal. Only two species of goatfishes commonly occur in the Caribbean, both of which are found in Anguilla.

Common Name – **Spotted Goatfish**

Scientific Name – *Pseudupeneus maculatus*

Local Name – Goatfish

Visual ID – White with row of three dark patches on body and two barbels under the chin. Can change colour dramatically, becoming dark mottled red when resting on sea floor (bottom), or showing a mixture of mottling with dark spots still showing (middle). Maximum size c.25 cm.

Abundance, Distribution & Behaviour – Common on most reef habitats, although prefer those closer to sandy areas. Usually seen in small groups, using their barbels to search for food in sand and rubble patches. Often accompanied by other opportunistic species. When not feeding they become inactive and lie on the bottom blending with the background. Juveniles are often seen on seagrass beds.

Common Name – **Yellow Goatfish**

Scientific Name – *Mulloidichthys martinicus*

Local Name – Queen Mullet

Visual ID – White with yellow mid-body stripe and caudal fin. Has two barbels which are usually folded under chin when not feeding, making it look very similar to the Yellowtail Snapper (page 132). Maximum size c.40 cm.

Abundance, Distribution & Behaviour – Common; can be abundant on some reef areas when they drift in large schools while not feeding. Seen in smaller groups while searching in sand and rubble patches for food with their barbels. Often accompanied by other opportunistic species.

Family – Sciaenidae: Drums

Drums derive their name from a low-pitched resonating sound made in the muscles surrounding their swim bladders. Most species, especially the juveniles, have elegant, elongated fore-dorsal fins, with striking black and white markings. The croakers also belong to this Family, with the **Striped Croaker** (*Bairdiella sanctaelucise* – not pictured) having been recently confirmed as present around Pelican Point.

Common Name – **Highhat**

Scientific Name – *Pareques acuminatus*

Local Name – Spade

Visual ID – Body is covered with thin black and white stripes, and has an overall dusky appearance. Adults are a more traditional fish shape than the following two drum species; all phases are easily distinguishable from these due to the bands present on the head (pictured). Maximum size c.20 cm.

Abundance, Distribution & Behaviour – Uncommon, preferring secluded areas of reef, often hiding under ledges or in protective recesses. Have been sighted among Elkhorn Coral colonies around Blowing Rock. Juveniles can occasionally be found in and around discarded concrete blocks on sandy areas or seagrass beds.

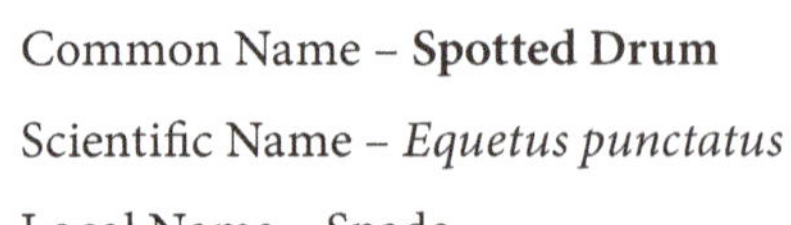
Photo Credit Ellen Muller

Photo Credit Gert-Jan Vandenbos

Common Name – **Spotted Drum**

Scientific Name – *Equetus punctatus*

Local Name – Spade

Visual ID – Body is covered in black and white stripes, while head has series of almost vertical black and white bars. Adult has black rear dorsal and caudal fin covered in white spots (top) that are absent in younger individuals (bottom) and helps to differentiate it from the similar Jackknife Fish (following). Juvenile has black dot on nose; the dorsal fin is extremely long and elegant, and reduces in size with age. Maximum size c.25 cm.

Abundance, Distribution & Behaviour – Uncommon, preferring secluded areas of reef, often hiding under ledges or in protective recesses. Juveniles can sometimes be found in and around discarded concrete blocks on sandy areas or seagrass beds.

Photo Credit Gert-Jan Vandenbos

Common Name – **Jackknife Fish**

Scientific Name – *Equetus lanceolatus*

Local Name – Spade

Visual ID – Black stripe extends over white body from tip of dorsal fin to end of caudal fin. The head has two almost vertical black bars; closely resembles the Spotted Drum (previous) but with fewer stripes and adult lacks white spots on fins. Juvenile (not pictured) lacks black spot on nose and can have a yellow wash and slightly shorter dorsal fin. Maximum size c.20 cm.

Abundance, Distribution & Behaviour – Uncommon, preferring secluded areas of reef, often hiding under ledges or in protective recesses. Juveniles can occasionally be found in and around discarded concrete blocks on sandy areas or seagrass beds.

Family – Echeneidae: Remoras

Remoras, also commonly known as 'sharksuckers', are a strange Family of fish which spend most of their lives hitching a ride with sharks, turtles, and even boats or divers! They attach themselves via a sucker pad located on the top of their heads, which is in fact a modified front dorsal fin. The Family name-sake, *Remora remora*, is a likely visitor to Anguilla as it is present in warm waters worldwide, but is not usually seen close to reef areas as it is more pelagic in nature. *Echeneis naucrates* (below), on the other hand, is the most common free-swimming remora and has been seen on a number of occasions looking very amorously at snorkellers! If attachment does occur the grip may be firm but it is harmless, and pushing the fish on its nose will encourage it to move on. Another species, the **Whitefin Sharksucker** (*Echeneis neucratoides*), has been sighted in Anguillian waters but is considered rare (see Spotted Eagle Ray picture – page 203). This species is distinguished by a wider caudal fin border.

Common Name – **Sharksucker**

Scientific Name – *Echeneis naucrates*

Local Name – Suckfish

Visual ID – Grey body with dark, white-edged mid-body stripe, and white borders on caudal fin when young. Has an oval sucker on top of head. Begins to lose markings with age and often only has short white stripe remaining on head. Maximum size c.1 m.

Abundance, Distribution & Behaviour – Uncommon while free-swimming, although has been sighted over seagrass beds at Little Bay and Rendezvous Bay, presumably looking for Green Turtles to hitch a ride with. Usually sighted when attached to sharks, rays and turtles although host abundance is normally low, so again, sightings are uncommon. When encountered by a diver the free-swimming Sharksucker becomes extremely amorous, allowing itself to be readily touched while it attempts to attach.

Are remoras parasites? Although remoras look like parasites when they hitch a ride from a larger species, they do not in fact harm the creature (aside from adding a tiny bit of drag), and so are not considered parasitic. The relationship between driver and passenger is more symbiotic in nature, where the remora benefits from a free ride and detaches to scavenge food debris left by the host when feeding. The water movement also helps ventilate the remoras gills, and the host's size offers protection. In return it is reported that the remora helps clean the host of skin parasites. Interestingly, fishers in the Indian Ocean and off the coast of Africa use remoras to help catch fish and/or turtles. In these regions remoras are caught and kept alive with a line attached. When a large fish or turtle is spotted, the remora is released into the water and heads straight towards its potential host. If small enough, this host can be hauled directly into the fishing vessel, whereas larger individuals are harpooned once they have been dragged within range. Happily enough for all species involved, this practice is not reported to occur in the Caribbean.

Eels

Family – Muraenidae: Moray Eels

Morays, often referred to as moray eels, have one fin beginning behind the head, forming a long, continuous frill which encircles the tail and extends mid-way down the belly. They also lack scales, instead being coated with a clear layer of mucus. They are often seen protruding from a crevice while menacingly opening and closing their mouths; this is not an aggressive action, but in fact aids respiration by circulating water through their gills. Morays are timid creatures and will usually retreat deeper into the reef if approached. Having said this, they have very powerful jaws and sharp teeth, so it is better to leave them unmolested in case a finger is mistaken for a tasty meal. Most species are nocturnal, leaving their dark recess at night to forage, although other species, for example the Spotted Moray and the Goldentail Moray, feed during the day, often in association with groupers and other fishes.

Common Name – **Green Moray**

Scientific Name – *Gymnothorax funebris*

Local Name – Green Moray

Visual ID – Green with heavy body and no distinctive markings. Maximum size c.2.5 m.

Abundance, Distribution & Behaviour – Rare. Spends most of the day hiding in reef recesses so is easily overlooked. Head is usually seen extending from recess opening. Known to frequent ship wrecks.

Common Name – **Spotted Moray**

Scientific Name – *Gymnothorax moringa*

Local Name – Conger Eel

Visual ID – White with dark speckles although may have yellowish tinge and so resemble the Goldentail Moray (next), especially when its body is hidden. A white iris is the most reliable way to differentiate this species. Maximum size c.1 m.

Abundance, Distribution & Behaviour – Occasional, hiding in reef recesses or among rubble during the day, usually with only its head showing. May also be found on seagrass beds and is probably the most common moray in Anguillian waters.

Common Name – **Goldentail Moray**

Scientific Name – *Gymnothorax miliaris*

Local Name – Conger Eel

Visual ID – Dark to brown body with yellow-gold speckles. Speckles become dense towards tail. Somewhat similar to the Spotted Moray (previous), especially when its body is hidden; a yellow ring around pupil is most reliable way to differentiate this species. Maximum size c.60 cm.

Abundance, Distribution & Behaviour – Uncommon. Hides in reef recesses during the day with only its head showing.

Common Name – **Chain Moray**

Scientific Name – *Echidna catenata*

Local Name – Conger Eel

Visual ID – Blackish brown body with irregular, often chain-like, bright yellow bars. Maximum size c.75 cm.

Abundance, Distribution & Behaviour – Occasional. Hides in reef recesses during the day with only its head showing.

Family – Ophichthidae: Snake Eels

Snake eels have a similar continuous fin to the morays, but differ by the presence of a small pectoral fin. These fins are often so reduced that they are hard to see, thus giving them the appearance of a sea snake; sea snakes are not present in Caribbean waters. Generally snake eels are reclusive and shy, hiding in reef recesses or burrowing under the sand during the day. Occasionally, however, they may be seen in daylight.

Common Name – **Sharptail Eel**

Scientific Name – *Myrichthys breviceps*

Local Name – Sea snake

Visual ID – Grey with small yellow spots on head and larger diffused paler yellow spots down rest of body. Maximum size c.1 m.

Abundance, Distribution & Behaviour – Uncommon. Can be found on many different habitats from shallow reefs to seagrass beds. Has been seen free-swimming by day on both Limestone Bay reef and Little Bay seagrass beds; is more active at night when it hunts. Sometimes buries itself and moves within sand.

Common Name – **Goldspotted Eel**

Scientific Name – *Myrichthys ocellatus*

Local Name – Sea snake

Visual ID – Light brown body with yellow spots that have diffused black borders. Maximum size c.1 m.

Abundance, Distribution & Behaviour – Rare, usually hiding during the day in sand or within reef structure. Is able to move beneath the sand.

Family – Congridae: Conger Eels

Conger eels, despite their foreboding name, are only represented by a very small and nervous species in the Caribbean. They have a similar continuous fin to the morays but differ, like the snake eels, by having a small pectoral fin. Only one species has been confirmed in Anguillian waters; other species may exist here but are unlikely to be seen as they usually hide deep within the reef structure.

Common Name – **Brown Garden Eel**

Scientific Name – *Heteroconger longissimus*

Local Name – None

Visual ID – Thin dark brown to grey body with slightly protruding lower jaw. Maximum size c.45 cm.

Abundance, Distribution & Behaviour – Locally abundant. Live in sand burrows from which they extend their head and upper body, facing the current, to catch passing plankton. They often sway around like plants in a breeze hence the reference in their name to a garden. One such garden can be found close to the wreck of the M.V. Oosterdiep (bottom image).

Sharks and Rays

Family – Rhincodontidae: Carpet Sharks

Carpet sharks are distinguished by their small eyes which are set well back on the head and a large upper lobe to the caudal fin, with a small, often indistinct, lower lobe. Carpet sharks are not considered a threat as most species either have small, almost sandpaper-like teeth used to grind shellfish or they are filter feeders – such as the largest fish in the world, the **Whale Shark** (*Rhincodon typus*). This latter species has been reported in Anguillian waters, although it is considered rare and as such extremely unlikely to be encountered while diving or snorkelling.

Common Name – **Nurse Shark**

Scientific Name – *Ginglymostoma cirratum*

Local Name – Nurse Shark

Visual ID – Dark grey with two barbels on upper lip and two, almost equally-sized, dorsal fins set well back. Caudal fin has no distinct lower lobe. Maximum size c.4 m.

Abundance, Distribution & Behaviour – Occasional. Can be found in almost any habitat from shallow lagoons to outer reefs. Usually inactive, laying in protected locations during the day. On one occasion three sizeable individuals were observed in a small cave-like recess at Katouche Bay. Not considered a threat although their small teeth may still produce a painful bite if molested.

Family – Carcharhinidae: Requiem Sharks

Requiem sharks, recognised by their pointed nose, are probably one of the most exciting creatures to see while diving or snorkelling. Despite their reputation, most species are wary of divers and will usually swim away after little more than a second glance. The only real danger comes with spearfishing when blood in the water may trigger aggressive behaviour. Having said this, all species should be treated with the utmost respect; as with all wild animals their behaviour can be unpredictable. The three species listed below are those most likely to be seen in Anguillian waters. Reports of other species have been made over the years, for example, the **Bull Shark** (*Carcharhinus leucas*) but confirmation of their presence or distribution has yet to be made. Other species confirmed in Anguillian waters but not pictured in this book include the **Tiger Shark** (*Galeocerdo cuvier*) which is bluish grey with pale belly and dusky bars and blotches on body that often fade with age, and the **Blacktip Shark** (*Carcharhinus limbatus*) that is bluish silver-grey with a white belly, anal fin and black tipped pectoral fins that may fade with age. Neither are common, athough the Tiger Shark has been sighted close to the sea rocks along Island Ridge, and occasionally larger individuals are landed by fishers – especially in the Crocus Bay and Sandy Ground areas. All sharks can be aggressive, especially in the vicinity of spearfishing, but the Tiger Shark is probably the most dangerous in Anguillian waters.

Photo Credit Stuart Wynne

Common Name – **Reef Shark**

Scientific Name – *Carcharhinus perezii*

Local Name – Reef Shark

Visual ID – Silver-grey with white belly. Anal fin is dark; dorsal fin begins behind pectoral fin. These latter two features differentiate it from the similar Blacktip Shark. Maximum size c.3 m.

Abundance, Distribution & Behaviour – Uncommon. Considered occasional in some locations, especially around Dog Island and some of the wrecks, but may be seen on most reef habitats. Pups can be found during May each year in the shallows at Shoal Bay East and Prickly Pear Cay. Considered dangerous especially in the vicinity of spearfishing.

Family – Sphyrnidae: Hammerhead Sharks

Hammerheads are the most recognisable Family of sharks due to their distinctive hammer-shaped head, which is flattened and extends horizontally. The exact purpose of this head shape is unknown, but as their eyes are on either end of this 'hammer' it is believed to aid sensory perception – potentially increasing the shark's ability to judge distances: Other suggestions to its function include manoeuvrability and manipulation of prey. Sightings of hammerheads have been made over recent years in Meads Bay although identification to species was not made. The **Scalloped Hammerhead** (*Sphyrna lewini*) is reportedly the most common species in the Caribbean and is considered dangerous to humans, especially in the vicinity of spearfishing.

Family – Dasyatidae: Stingrays

Stingrays are closely related to other similar-looking fish Families such as the electric rays, skates and guitarfishes. All are bottom-dwellers that have greatly enlarged pectoral fins giving them a disc-like shape, but none of these latter Families have been sighted in Anguillian waters and only one species of stingray has been identified to date. The mouth is on the underside and they feed on molluscs and crustaceans which they find in the sand. Their technique, once prey has been located by smell, is to pound the sand with their head, thus uncovering their meal. This leaves small pits in the sand which are a tell-tail sign of a ray's presence in the area. Often a feeding stingray will be accompanied by opportunistic fish such as the Bar Jack (page 110), Pluma (page 120) or Spotted Goatfish (page 194) which hover behind the hunter, darting in to grab any overlooked food. When not feeding the stingray rests motionless on the bottom having partially covered itself with sand.

Photo Credit Stuart Wynne

Common Name – **Southern Stingray**

Scientific Name – *Dasyatis americana*

Local Name – Stingray

Visual ID – Brownish grey to black with slightly pointed tips of wings. Has whip-like tail with one or two venomous spikes. Maximum size (excluding tail) c.1.5 m.

Abundance, Distribution & Behaviour – Common to occasional depending on habitat, preferring sandy areas where they hunt for their buried food. Often lie motionless on the bottom partially covered in sand. Particularly common around parts of Prickly Pear Cay. To avoid being spiked by tail it is best not to molest them. When walking in shallow water where they may be present, shuffling ones feet, rather than taking steps, will scare them off before they can be startled.

Family – Myliobatidae: Eagle Rays

Eagle rays are a very majestic Family that swims by flapping their fins in a similar way to that of a bird flying. Unlike the stingrays they are not bottom-dwellers and swim freely, often in pairs. They are represented by only one species in the Caribbean.

Common Name – **Spotted Eagle Ray**

Scientific Name – *Aetobatus narinari*

Local Name – Eagle Ray

Visual ID – Dark upper body with white spots and pale belly. The head is somewhat pronounced with pointed snout; tail long and thin. Maximum size (wing-tip to wing-tip) c.2 m.

Abundance, Distribution & Behaviour – Uncommon to occasional. Cruises reefs and sand flats and are often seen in pairs. Usually sighted in areas closer to open ocean, such as Sandy Island, Madeariman Reef and along the south coast; is also sometimes seen in very shallow water. A clue to its presence is the small sand cloud it leaves as evidence after nosing out molluscs in the sand. Note the Whitefin Sharksucker (page 197) attached to its ventral side.

Family – Mobulidae: Manta Rays

Manta rays are the largest of all rays and can reach over six metres in diameter. They have a characteristic head shape that is used while swimming to funnel plankton and other small swimming creatures into their mouths. If they find particularly rich feeding grounds mantas can congregate in large numbers looping-the-loop gracefully in the water until they have had their fill. One species is known in the Caribbean, the Giant Manta (*Manta birostris*), although it is considered oceanic and only rarely seen close to reefs.

Common Name – **Giant Manta**

Scientific Name – *Manta birostris*

Local Name – Manta Ray

Visual ID – Dark upper body with white belly that often has dark blotches. Has two long protrusions either side of mouth that help 'funnel' plankton when feeding. Maximum size (wing-tip to wing-tip) c.5 m.

Abundance, Distribution & Behaviour - Rare, and primarily oceanic. Will swim, often breaching the surface, while feeding on plankton. Two juveniles were sighted near the jetty at Crocus Bay in 2015.

4.2 Class – Mammalia: Mammals

Order – Cetacea: Marine Mammals

Cetaceans are a relatively small group of mammals containing fewer than 100 species, the vast majority of which live in salt water. They are adapted to life in the ocean better than any other group of mammals and in Anguillian waters are represented by dolphins and whales. Only a small number of species have been included in this book as the complete list is, so far, unknown and those present are unfortunately very rarely seen while diving or snorkelling. Most sightings are made by boat when a 'breach' occurs, but the rapid nature of such an encounter, combined with limited visual contact (the breach probably occurring at a distance or only involving a small portion of the animal), means identification is only tentative at best. Furthermore, many species may only be transient visitors, passing through Anguillian waters during their annual migratory routes and so are very unlikely to be encountered at all. The three species of dolphin listed below are the only ones to have been positively identified to date and the two species of whale listed are those considered most likely to be present. The Department of Fisheries and Marine Resources is very keen to hear reports of any cetaceans sighted in surrounding waters.

Family – Delphinidae: Oceanic Dolphins

Oceanic dolphins are the largest and most diverse Family of cetaceans which contains all 'true' dolphins, blackfish (i.e. Killer Whales) and some coastal and partly riverine species. Many of these species migrate long distances and have a range incorporating the Caribbean, so many more species than those listed below might be sighted in Anguillian waters. For example, in November 2003 thirty six **Short-finned Pilot Whales** (*Globicephala macrorhynchus*) were found stranded on a beach in St. Martin and sporadic, unconfirmed sightings of other unidentified species have been made. The two oceanic dolphins featured are those confirmed to be sighted on a relatively regular basis. Dolphins are social creatures forming pods of varying sizes, usually containing a family unit of relatives. They classically 'speak' or 'sing' to one another, a behaviour which over the years is gaining credibility as a somewhat complex language. Dolphins do not remain in one place and travel from area to area often in seasonal migratory patterns. Other than these two species, the **Striped Dolphin** (*S. coeruleoalba*) has also been confirmed in Anguillian waters when, in 2008, numerous pods were seen migrating through the channel between Anguillita and Dog Island. Some of these pods were also of *S. frontalis* (page 205).

Common Name – **Bottlenose Dolphin**

Scientific Name – *Tursiops truncatus*

Local Name – Dolphin/Porpoise

Visual ID – Bluish grey body with off-white, light grey or pinkish underside. Colourations are varied and complex but basically form most people's idea of the classic dolphin; they are generally the species used in water parks and movies. Size and shape also varies from one individual to another. No distinctive features overall.

Abundance, Distribution & Behaviour – Unknown. Have been sighted in deeper water all around the island, the offshore cays and recently in shallower water at Flat Cap Point. Seasonality is unknown. Individuals form pods, usually consisting of a family unit.

Common Name – **Atlantic Spotted Dolphin**

Scientific Name – *Stenella frontalis*

Local Name – Dolphin/Porpoise

Visual ID – Grey to blue dorsal side and whitish ventral side. Has a light covering of spots which are more visible on paler parts of body than dark areas. Dorsal fin is usually unspotted. There are great variations between spot patterns which taxonomists have puzzled over for years, but it is now accepted that the Atlantic Spotted Dolphin is a separate species. An adult is usually between 1.5 and 2 m in length.

Abundance, Distribution & Behaviour – Unknown, although as a migratory species their presence will probably be seasonal. A large number of pods, containing both *S. frontalis* and *S. coeruleoalba* (page 204) were seen in June 2008 in the channel between Anguillita and Dog Island. Also often seen around Prickly Pear Cays.

Family – Physeteridae: Sperm Whales

Sperm whales are probably the easiest Family of whales to identify because of their bulbous head and narrow, underslung lower jaw. The head contains a huge cavity filled with a milky-white, waxy substance known as spermaceti. Interestingly, the Sperm Whale's name comes from the misidentification of this substance as sperm. The spermaceti's function is not fully understood but it is believed to help control buoyancy and/or to act as an acoustic lens to aid in echolocation. Each whale contains up to three tons of spermaceti which years ago was considered a very valuable substance, being used to make candles, soap, cosmetics and machine oil. The whale was also valued for its blubber, sperm oil and ambergris (a fixative once used in the perfume industry). Because of its value, sperm whales were hunted close to extinction during the last century and are still considered vulnerable by the IUCN (International Union for Conservation of Nature).

Common Name – **Great Sperm Whale**

Scientific Name – *Physeter macrocephalus*

Local Name – Whale

Visual ID – Dark body with off-white underside. Huge 'forehead' with tiny eyes and narrow, barely visible lower jaw. Older individuals usually scarred by encounters with Giant Squid which they hunt for food. Adult size ranges between 11 and 18 m.

Abundance, Distribution & Behaviour – Unknown; sightings have been made from offshore fishing grounds well away from reef areas. The species usually form two main groupings: 'bachelor schools', consisting of young sexually inactive males; and 'breeding schools', consisting of females with young of both sexes. Groups contain approximately two dozen animals; in rare cases hundreds, or even thousands, have been reported together (not in Anguilla!). Older males are usually solitary or in small groups of around half a dozen, and spend a few hours at a time with the breeding group during breeding season. Although the most heavily exploited species of whale, the Great Sperm Whale is still relatively common throughout the world's oceans.

Family – Balaenopteridae: Rorqual Whales

Rorqual whales are the largest whales found in the worlds oceans, most of which migrate long distances between their warmer waters breeding grounds (winter months) and colder water feeding grounds (summer months). The Family has suffered from intensive exploitation through hunting and many populations are still severely limited or have disappeared altogether. Many more species than the one listed below probably visit Anguillian waters at various times of the year, but confirmation has yet to be made.

Common Name – Humpback Whale

Scientific Name – *Megaptera novaeangliae*

Local Name – Whale

Visual ID – Dark body with white flippers. These elongated flippers, together with the knobbly head, make it one of the easiest whale species to identify. Size can range from 11 to 15 m.

Abundance, Distribution & Behaviour – Unknown, although it will probably only be in the Caribbean's warmer waters during the winter months when breeding. A breeding population is known to exist near Puerto Rico but currently no such group has been confirmed in our waters. Sightings have been reported at offshore fishing grounds well away from reef areas.

Are there seals in the Caribbean? The **Caribbean Monk Seal** (*Monachus tropicalis*), the only known seal species to be native to the region, is now reported to have been hunted to extinction. The last sighting was of a small colony on Serranilla Bank, a collection of tiny cays and shallow reefs between Honduras and Jamaica, in 1952. The species was not officially declared as extinct until 2008 because over the years there have been sporadic reports of sightings by fishers and divers in Haiti and Jamaica. Recent scientific expeditions however failed to find any specimens. It is likely the creatures sighted were, in fact, wandering Hooded Seals (*Cystophora cristata*), a species normally living much farther north. Although no historic records exist documenting the presence of Caribbean Monk Seals here in Anguilla, it is possible that they once frequented our waters, as suggested by the name of Seal Island, located next to Prickly Pear Cays.

4.3 Class – Reptilia: Reptiles

Order – Chelonia: Turtles and Tortoises

Within the above classification, the Superfamily Chelonioidae contains the most well known marine reptiles, the sea turtles, a majestic group of ocean-going travellers that only come ashore to lay their eggs. All sea turtle species are threatened to one degree or another and should be treated with due respect. It is currently illegal to catch or be found in possession of either a living or dead individual in Anguilla. Adult sea turtles migrate, although migration patterns vary between species and even between individuals of the same species. Two species, the Green & the Hawksbill Turtle, are commonly seen here as juveniles, spending their younger years foraging either on reefs or seagrass beds, regularly coming up and resting on the surface to breathe. The Leatherback turtle is also found here, and like the previous two species, it comes up onto selected beaches to nest, although it is unlikely to be seen while diving or snorkelling. The Loggerhead Turtle has been sighted in Anguillian waters but is thought to be far less common, with no confirmed nesting activity reported.

Although protected by law in Anguilla, turtles continue to be threatened by more indirect means. Marine garbage can be inadvertently swallowed by feeding turtles and block their digestive systems. Entanglement, followed by drowning, can be caused by old discarded or lost fishing gear. Coastal development is another source of stress for turtles, where bright lights or loud noises disturb nesting individuals. If females are disturbed continuously, and as a result do not successfully build their nests, they may end up dropping their eggs out to sea. Hatchling babies can also be disorientated by these disturbances, and become lost before they reach the ocean. Removal of beach flora does not only lead to sand erosion, but also affects hatchling sex ratios. Nest temperature is crucial in controlling this, so a certain percentage of nests need to be built in shaded areas to ensure one sex isn't over represented.

For these reasons adult turtles remain endangered and, although data are still being collected, it is suspected that nesting frequency is still in decline on Anguilla. If a nesting female is sighted people are requested not to disturb her; instead contact the Department of Fisheries and Marine Resources or the Anguilla National Trust who both monitor populations.

Family – Dermochelyidae: Leatherback Turtles

Leatherback turtles, a Family that consists of only one species worldwide, can be easily distinguished by the lack of a bony shell, instead having ridged leathery skin and oily flesh. It has the widest distribution of all turtle species, largely because it can regulate its body temperature to a certain extent, thus allowing it to visit more temperate areas. It is also the largest turtle species and can dive deeper than any other, having been documented at depths of over a kilometre. This species is currently considered critically endangered.

Common Name – Leatherback Turtle

Scientific Name – *Dermochelys coriacea*

Local Name – Leatherback Turtle

Visual ID – Predominantly black and easily distinguished by a carapace that lacks hard plates, instead having tough leather-like skin with vertical ridges running down back. The largest species of turtle, growing up to 2 m in length.

Abundance, Distribution & Behaviour – Although not often seen in water because it spends most of its time living a pelagic existence while hunting for its favourite food (jellyfish), it is a relatively common visitor to select beaches in Anguilla where it lays its eggs, usually at night. Over recent years the most popular beach known for Leatherback nesting is Captains Bay.

Family – Cheloniidae: **Hardback Turtles**

Hardback turtles comprise all species of marine turtles apart from the Leatherback Turtle. Hardback turtles, as their name suggests, possess a hard carapace made up of several bony plates known as scutes. Adults of the species are rarely sighted while diving; most encounters are with juveniles that spend their time foraging in near-shore areas until they reach maturity and embark on a more migratory lifestyle. All species of hardback turtles are considered endangered, with the Hawksbill listed under the greatest threat (listed as critically endangered).

Common Name – **Green Turtle**

Scientific Name – *Chelonia mydas*

Local Name – Greenback Turtle

Visual ID – Carapace scutes are less patterned than those of the Hawksbill turtle and a paler brownish-tan colour. Edges of scutes are well defined as they do not overgrow each other, and the rear margin is not noticeably jagged. Has two small plates between the eyes, again distinguishing it from the Hawksbill (previous) – see diagram. Beak is not noticeably hooked.

Abundance, Distribution & Behaviour – Juveniles are common in certain coastal regions, especially seagrass beds where they forage for food. Nesting females are considered uncommon. Juveniles are also sometimes sighted on reef areas. Usually most active during sunrise and sunset.

Common Name – **Hawksbill Turtle**

Scientific Name – *Eretmochelys imbricata*

Local Name – Hawksbill Turtle

Visual ID – Carapace scutes are patterned brown and tan, overgrowing those directly behind. The rear part of the carapace margin is often noticeably jagged, and the head has four small plates (two pairs) between the eyes which can help to distinguish it from the Green Turtle (next) – see diagram. Its 'beak', often noticeably hooked, is used to feed on sponges or scrape other food stuffs from rocks.

Abundance, Distribution & Behaviour – Juveniles are common in certain coastal regions with nesting females considered occasional. Juveniles forage on rocky reef areas but are also occasionally sighted on seagrass beds. Usually most active during sunrise and sunset.

Notes on other sea turtles: Of the four species found worldwide that are not listed in this book, only one, the Flatback Turtle (*Natator depressus*) is technically impossible to see in the Caribbean region, as it lives exclusively around the coast of Australia. The remaining three are found in the Caribbean, although their distribution varies: the **Kemp's Ridley Turtle** (*Lepidochelys kempii*) is most likely to be seen near the coast of North or South America; whereas the **Olive Ridley Turtle** (*Lepidochelys olivacea*) is a more frequent visitor to the islands although still rare in most areas; the **Loggerhead Turtle** (*Caretta caretta*) is probably the most common of the three species, although only anecdotal sightings have been made in Anguilla, and no nests confirmed. It is similar in appearance to both Green and Hawksbill turtles (previous two species), but is generally bulkier with a thick neck and domed carapace. It can be reliably differentiated by five pairs of lateral scutes, rather than the four that these other species possess.

Chapter 5 - Artifacts

This chapter is dedicated to those items of biological origin that do not fall into the previous category chapters of this book. These items may be observed either underwater or washed up on beaches, with some being quite common and of local origin, while others are rarer, sometimes even arriving in Anguillian waters after travelling long distances with ocean currents. Many of these items are internal remnants of creatures, or from those that have very cryptic, often infaunal (under the sand or mud), life histories, such as many of the sea shells. Also featured are those creatures whose appearance drastically changes after death, such as the echinoderms. Although Anguilla does not have huge densities of shells that wash up on the beach daily, the variety is still such that only a selection of reference species has been possible to include.

This chapter is not intended to encourage the collection of biological artifacts from Anguilla's beaches as the majority are mineral based and naturally break-down over time, either contributing particulate matter to beach formation or dissolved minerals to water chemistry. This serves an important ecological role in ecosystem integrity and as such the Anguilla Government does not condone the removal of any natural materials from the beaches, with their removal from Marine Park areas prohibited by law.

Finally it is appropriate here to make one of the most important pleas to those reading this book. Although beach combing can be an interesting and sometimes rewarding hobby for many people, PLEASE do not add to the artifacts found on the beach by leaving garbage behind. Not only is it an eyesore, but it can also be harmful to the life we are celebrating within this book. For example, turtles can mistake plastic bags, cigarette butts, and other items as food which, if ingested, may block their digestive system. It is especially important for boat owners to be aware of this as their garbage can blow around and end up in the sea even if they didn't mean it to. The majority of rubbish found on beaches around the world actually originates from the ocean rather than beach goers, and so boat owners play a critical role in reducing its impact. The Anguillian National Trust organise beach cleanups from time to time, but otherwise it is down to hotel developments and the general public to keep these areas clean. One fate for all this garbage is to become part of one of the many 'sculptures' that appear from time to time (see inset image). While we admire the creativity, it doesn't permanently remove the material out of harms way. The most useful contribution a single person can make is to collect any rubbish they see in their vicinity when they leave an area of beach, and in this way the continual tide of material brought up by the sea, combined with that left by inconsiderate beach goers, may be kept in check.

Sculpture of beach artifacts found on Savannah Bay (Stuart Wynne)

5.1 Molluscan Remains

The group of creatures that contributes the majority of artifacts to Anguillian beaches is by far the molluscs, but the diversity of species is such that only the most common will be generically covered here.

Olive shells

Coffee Bean Trivia

Flamingo tongues

External Molluscan Remains

Olive shells (*Oliva* sp.) and **cowries** (*Cypraea* sp.) are more commonly seen washed up on a beach than they are alive underwater. They are prized among shell collectors for their lustrous and beautifully patterned exterior, although in some parts of the world over-collection has lead to greatly reduced numbers. This lustrous finish is produced by their mantle, a fleshy protrusion that covers most of their outer surface while alive underwater. After death, with the mantle retracted, their appearance changes significantly, but they will retain their lustre for a week or two until the sea erodes it away. Some species, especially the Olive Shells, live partially under the sand where they forage for food, which adds to their already cryptic nature.

The **Coffee Bean Trivia** (*Pusula pediculus* - formerly *Trivia pediculus*), is a related species, looking very much like a tiny cowrie with a series of ridges on its surface. It lacks an external mantle and thus appears less polished than many of its relatives. Although it does not have the camouflaging mantle and does not live an infaunal life, its small size means it is almost exclusively seen only when washed up on a beach after death.

Flamingo Tongues (*Cyphoma gibbosum*) although more distantly related, bear some resemblance to cowries, and have a similar external mantle that renders the shell surface shiny immediately after death. This species and other members of the Genus *Cyphoma* (page 89) have been included here as when washed up on the beach they appear very different from how they look underwater. Their mantle is usually cream coloured and covered with orange ocellated spots, a pattern that differs between species. Once the mantle is absent, the shell that remains is uniformly white, only bearing a shape resemblance to the living creature.

Often not resembling live creatures, the fragments of larger molluscs can be found in quite large piles having been left by fishers once they have removed the edible creature from inside. Of special mention here are the **Queen Conch** (*Lobatus gigas* - page 85) & the **West Indian Topshell** (*Cittarium pica* - page 86); the latter of which is known locally as a whelk. These are of special note because when fresh their internal surfaces are of significant beauty, with the conch a powerfully vibrant pink colour and the whelk having a rainbow-like mother of pearl lustre. Such piles, known as 'middens', are especially noticeable at Sandy Ground.

Conch midden ay Sandy Ground

Queen Conch (page 85) remains

Queen Conch (page 85) remains

West Indian Topshell (page 86) remains

Another contributor to molluscan remains belongs to those left by the **chitons** (page 91). Although they can sometimes be found on the beach whole, curling up as their connective tissues dry, they often break apart leaving behind only scattered plates. On beaches with nearby rocky areas covered in chitons these plates can by very common.

Fuzzy Chiton whole remains

Fuzzy Chiton individual shell plates

Worm Snails (*Siliquaria* sp.), although very similar in appearance to tubeworms (page 72 & 217), are in fact gastropods that do not coil their shell, or only do so partially, creating a small spiral at their rear end. They either live in colonies, or attach themselves singly to rocks and shells. Although not considered particularly common, at times they are found on beaches in relatively large numbers.

Worm snail remains

Worm snail remains

The **Tusk Shells** belong to a group known as the Dentaliidae which contains over 1000 living species belonging to approximately 14 Genera. They can inhabit both shallow and deep waters where they live burrowed into sandy bottoms filter feeding on tiny planktonic creatures and micro-organisms. When washed up on the beach their shells look like tiny tusks (pictured left). Of the different species many appear very similar, making species and even Genus identification difficult.

While a number of bivalve shells can be found washed up on the beach, one of the most beautiful is probably the **Sunrise Tellin** (*Tellina radiata*) although to find an intact specimen is rather rare due to how thin and delicate their 'valves' are. Being infaunal in nature it lives buried in the sand, only being exposed after death if natural sand movement sifts it to the surface. Other potential bivalve finds are pictured below.

Sunrise Tellin (*Tellina radiata*)

Speckled Tellin (*Tellina listeri*)

Angulate Tellin (*Tellina angulosa*)

Zebra Arc Shells (*Arca zebra*)

File Clam (*Lima* sp.)

Calcio Clam (*Macrocallista maculata*)

Internal Molluscan Remains

The delicate-looking **Rams Horn Shell** (*Spirula spirula*) are often found washed up on Anguilla's beaches and are, surprisingly, the internal shell of a deep-water squid. This rarely seen squid is the only living member of its Genus. The coiled shell is delicate, light-weight and very buoyant, thus after the squid has died and rotted, its internal coil floats up to the surface and can drift vast distances on ocean currents. Its spiral shape is divided into separate gas-filled chambers, thus if part of the shell breaks it will retain most of its buoyancy. This design is somewhat reminiscent of the Nautilus (not featured in this book), that itself is the closest living relative to the Ammonites, an extinct group of marine creatures that dominated the oceans millennia ago. Probably due to prevailing currents Rams Horn Shells are most commonly found on the beaches along the south eastern coast.

Rams Horn Shell

Cuttlefish pens, as their name suggest, belong to cuttlefish. They are in fact the internal bone that gives this creature, a relative of squid and octopus, a structural base while also being used for buoyancy control. For this reason when the cuttlefish dies the internal bone floats to the surface and can drift great distances on ocean currents. In fact, there are no cuttlefish living in the Caribbean region and thus all cuttlefish pens found have floated in from North America, South America, or even Europe. This travel distance is apparent as the pens are often found with **Gooseneck Barnacles** (*Lepas anatifera* – page 83) growing on them, and are always somewhat brittle which suggests they have spent some time at sea. These structures are most commonly associated with bird cages where they are hung to provide the bird with a useful mineral source. Probably due to prevailing currents cuttlefish pens are most commonly found on the beaches along the south eastern coast.

Partially eroded cuttlefish pen

The wonderful detail that exists underwater will never cease to astound the interested observer. Intricate plant structures (*Caulerpa verticillata* - page 22) and tiny coral detail (*Stylaster roseus* - page 41) are prime examples. Similarly, amazing details can also be found in many shell examples washed up on the beach. One such example that stands out from a perspective of beauty is that of the **Zebra Nerites** (*Puperita pupa*), with their intricate black and white patterns and lovely soft yellow interior. They are not particularly common, but can be found from time to time on certain beaches.

Notes on Other Molluscan Remains

Most gastropods (page 84-91) have a tough 'horny' structure attached to their foot known as the operculum. In Conchs, this operculum (plural operculae) is a hook-like shape and is used by the creature to drag itself along the sandy substrate. In other species, for example the **West Indian Top Snail** (*Cittarium pica* - page 86), the operculum is a somewhat flimsy flattened spiral that serves only as a deterrent to potential predators. In other Genera however they have become thick, domed, shell-like structures that wash up frequently on beaches, with their origin often a mystery to beach combers. Most of these operculum types belong to species of **Turban Shells** (*Turbo* sp. - below right) and in some parts of the world go by separate names such as Shivas Eye, Shivas Shell & Cats Eye, whose more vibrant colours make them popular jewellery items. Those types found in Anguilla are less dramatically coloured (below left), although they are still considered items worthy of collection by some. There are variations with the operculum design among the turbans, and other related Genera possess similar structures also, for example the star snails and the **Astrea Snail** (*Astraea tecta*), and the more distantly related Moon Shells (see following section).

Molluscan operculums

Turban Shell (*Turbo* sp.)

Moon Shells (below left) belong to a number of Genera that are all within the Family Naticidae. The exact species that live in Anguillian waters have yet to be determined, but all are almost spherical with a smooth outer surface that usually has a polished lustre. This finish is produced by the foot that often wraps itself almost entirely around the shell (in a similar way to the mantle of the cowries (page 210). Moon shells are carnivorous and live almost exclusively under the sand with the only sign of their presence often being sand collars (below right). These sand collars are crafted by the foot of the female, and consist of two layers of sand stuck together with mucus, in between which she lays her eggs. These frilly-looking wheel-like formations are very delicate and normally break apart once the eggs hatch.

Moon Shell

Moon Shell sand collar

5.2 Echinoderm Remains

Arguably one of the most fascinating group of artifacts found on beaches are those left by the Echinoderms (sea urchins, brittle stars, etc – page 95-103). The most common of these remains found in Anguilla are by far the tests of the domed urchins, who although very delicate wash up in sufficient numbers that whole specimens are very common, especially those belonging to the Rock-Boring Urchin (*Echinometra lucunter* – page 100). Beaches close to rocky areas where the urchins live have the highest density of specimens. The urchin tests are variable depending on species and so it is often interesting to take a close look at them to spot the differences.

Similarly whole specimens of the infaunal urchin groups can frequently be sighted in sandy areas after they have died and sifted to the surface by natural sand movement. The most common species to find are the **Red Heart Urchin** (*Meoma ventricosa* - page 101) and the **Six-Keyhole Sand Dollar** (*Leodia sexiesperforata* - page 102), although sightings are more common snorkelling or after a retreating tide (tides in Anguilla are often hardly noticeable) as their tests are very delicate and rarely survive wave action without sustaining any damage.

Of the other echinoderms the **Cushion Star** (*Oreaster reticulatus* - page 96), Anguilla's most common sea star, is rarely found stranded on beaches as it prefers to live in sheltered areas with mid-range depths. In certain locations after a storm however they do wash up, along with a wide range of other creatures in quite significant numbers. Species of brittle stars on the other hand can be found more frequently, although still not on a regular basis. The most common species to find is probably the **Circle Marked Brittle Star** (*Ophioderma cinereum* - page 98) who usually live in shallow rocky regions, hiding in cracks and crevices. The **Blunt Spined Brittle Star** (*Ophiocoma echinata* - page 97) is also a potential candidate, although this species is more commonly fragmented – probably because it lives in slightly deeper water and thus has further to travel via wave action before it is washed ashore.

Slate Pencil Urchin tests (page 101)

Rock Boring Urchin tests (page 100)

Red Heart Urchin tests (page 101)

Two different brittle star skeletons (page 97)

5.3 Other Artifacts

Coral fragments are extremely common on certain beaches in Anguilla, sometimes dominating the entire coast (below left) or forming berms out into the sea that have been piled up by hurricanes or other storms. Such formations are common along the south coast, especially between Blowing Point and Sile Bay, where wave action is usually at its strongest. Coral fragments can be distinguished from other rocks by the corallite structure that is usually still visible on the surface and can even be used to identify the species responsible for its creation. This structure persists even after long periods of weathering because it is present through the entire 'rock' having grown as described on page 51. Other coral fragments are much smaller, being the remains of solitary polyps, less massive colonial species (such as fire coral or staghorn coral) or small colonies that had yet to reach a large size before their demise (below right).

Large coral fragments dominate parts of Anguilla's Coastline

Small coral fragments can be found washed up on many of Anguilla's beaches

Soft corals also wash up on beaches in quite considerable numbers; and are often easily recognisable because of their distinctive structure (below left). Sometimes however when fragments wash up on the beach they can look very much like parts of tree branches, especially when only the holdfast of a sea fan is found. These fragments can be distinguished from wood by examining their internal structure, which for soft corals is most commonly made up of concentric layers of the horny substance called gorgonin (page 65). Calcareous algae (page 19-20) also washes up on beaches, sometimes almost whole (below right), but more often the leaf-like segments break apart and sometimes form large areas of course 'sand', such as in certain parts of Shoal Bay East.

Soft coral remains

Calcareous algae remains - *Halimeda* sp. (page 19-20)

Sponge remains

Sponges can be commonly found on beaches, although they are more abundant in certain areas than others, and are generally quite recognisable as they usually retain their general structure. Some however do vary considerably or are washed up in unrecognisable fragments. Such fragments can be identified as sponges by the presence of their spicules, which remain in place after the rest of the animal has decomposed. These spicules are responsible for keeping the sponge in its original form and can themselves be used by an expert to identify species, although for the more casual observer this is not possible. Some species have very brittle spicules and thus crumble quite readily, while others are more flexible, for example the **Branching Vase Sponge** (*Callyspongia vaginalis* – page 31), and are reminiscent of traditional washing sponges; where their original name 'sponge' came from.

Calcareous tubeworms (page 72) belong to a number of Genera that are all within the Family Serpulidae. They secrete a calcareous tube within which they live, although the most common species burrow into rock or coral and so their tubes are rarely seen and do not wash up on beaches. Other smaller encrusting varieties are those more often seen by beach combers. For example, these can be found in clumps (Sea Frost - *Filograna huxleyi* - page 74), or singly encrusting the exposed surfaces of shells, rocks and other hard substrates (see below). Although the species present in Anguilla have yet to be categorised, and their small size means they are easily overlooked, they are surprisingly abundant and commonly found if one is willing to get on the floor and closely inspect the material washed up by the sea.

Calcareous tubeworm on a seagrass frond

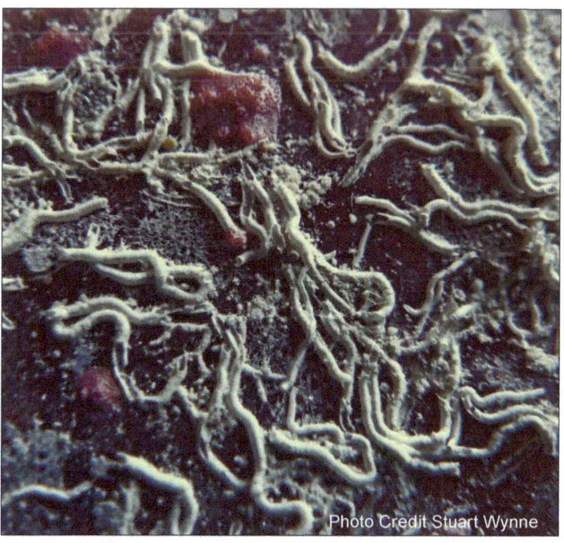

Calcareous tubeworms on a shells surface

Although not obvious at first glance because they resemble dark-coloured rock, **tar balls** that have been washed up on rocky shores deserve a mention here because of their biological origin. Their more direct origin however is from oil spills followed by weathering processes. When crude oil is weathered in the ocean the lighter fractions evaporate, leaving the heavier fractions behind. These then get broken up, usually into small coin-sized fragments that are heated by the sun and mixed with sea water. This ultimately produces a thick, sticky tar ball that may travel on ocean currents for thousands of miles before being washed up on a beach. The weathering process creates a crusty outside with a soft and gooey inside, that may be broken open by turbulence when making land fall, and becoming a nuisance to beach goers. Eventually, especially on rocky shores, the ball flattens and becomes encrusted to the rock although it will remain pliable for many years. Sometimes quite large tar balls can wash up, as can be seen in the photograph to the right taken at Corito Bay. Its large size does suggest minimal ocean transport, although as crude oil is not shipped to Anguilla its source remains a mystery.

Tar ball remains

On occasion hatched egg cases become dislodged from where they were originally fastened and wash up on the beach. Although these items are relatively rare in Anguilla, **Tulip Shell** egg cases (pictured right) is one example that may be found by keen beach combers. These pretty structures look like dried flower heads and are leathery and tough. Another example is the classic Mermaid's Purse that originates from the **Sting Ray** (page 202). Although many sharks and rays produce similar egg cases (known as oviparous), it is interesting to note that some species give birth to live young (viviparous), while others (such as the **Spotted Eagle Ray** - page 203) has eggs that hatch inside its body before the young are released (ovoviviparous).

Fossils, the final group in this chapter, are often found on beaches and all ultimately have a biological origin despite being mineral in nature now. In fact, Anguilla's geological history is such that fossils of marine creatures can be found all over the island, and not just on its coast. In fact, most of Anguilla's bedrock is limestone, which being of biological origin means Anguilla is essentially one big fossil! This limestone can be seen clearly in many coastal areas, such as around Sandy Hill Bay (pictured right). In some areas however the limestone is far softer or the rocks have other sedimentary origins. Sandstones are among

Tulip Shell (page 86) egg cases

Anguilla is essentially a large fossil formation

these and are made up of ancient beach deposits from when either sea levels were higher or rock strata lower. As these deposits were laid down, the creatures that lived and died at the time sometimes became buried and as subsequent layers were added, pressure built up and the sand became compressed. During this time mineral changes also occurred that rendered the compressed sand more rock-like, and fossilised the creatures that had become buried within it. This action can be clearly seen at North Cliffs (Katouche Bay) and Isaacs Cliffs (Between Long Bay and Sandy Ground). Here the surface layers easily crumble and the creatures within it are quite brittle, whereas the deeper layers are far more solid and the fossils robust. Although this process takes millions of years it is surprising how similar the fossilised creatures look compared with those alive today. Pictured to the right are photographs of a selection of the most common fossils likely to be found in Anguilla, some of whose relatives are featured elsewhere in this book.

Inflated Sea Biscuit (page 102) fossil relative

Fossil bivalve (left) and gastropod examples found in Anguilla

Fossil sea urchins found in Anguilla

Fossil gastropod found in Anguilla

Chapter 6
Supplemental Information

Presented in this chapter is information related to the threats faced by the marine environment, conservation efforts taking place on Anguilla, and suggested further reading and online resources. Over recent decades it has become apparent that the marine environment is more fragile than once thought and faces a number of threats that may affect its prosperity in the future. In the Caribbean for example a 70% decline in hard coral cover was been reported regionally to take place between 1970 and 2000. These are quite alarming figures, with photographic confirmation extremely disturbing (see below images). Without hard corals dominating the reef ecosystems, fewer new calcium carbonate structures are built to replace those lost through weathering. This weathering can occur once the coral colony that deposited the structure dies, or if damage occurs through wave action. Without these structures fish and other creatures do not have as many places to call home, and the storm protection that reefs afford beaches and other coastal areas is lost. Without reefs it is probable that Anguilla would have fewer beaches, and those that did persist would probably be narrow strips of sand that were constantly shifting, no longer resembling those that Anguilla is famous for today.

Elkhorn Coral on Seal Island Reef 1994 (John Bythel)

Finger Coral on Seal Island Reef 1994 (John Bythel)

Elkhorn Coral on Seal Island Reef 2015 (Stuart Wynne)

Finger Coral on Seal Island Reef 2009 (Stuart Wynne)

The reason for this decline in hard corals is complex and still not fully understood. It is part of a more widespread problem that affects both terrestrial and marine areas, known as habitat degradation. For coral reefs, such degradation is caused by a combination of factors that all work and interact with each other. Part of the complexity is a phenomenon known as a 'feedback loop', where, for example, a reduction in coral cover lowers habitat quality to the detriment of fish populations. With fewer herbivorous fish, algae can grow unchecked and outcompete corals, further reducing their cover and inhibiting new corals to settle and grow. Such loops can cause habitat health to spiral downwards.

6.1 Threats to the Marine Environment

The contributing threats can be interdependent and thus at times work in unison with each other, but have been split into categories and discussed broadly below:

Overfishing – Although this is a well documented phenomenon in many parts of the world, the extent and threat of overfishing in the Caribbean is still a matter of contention. For example, although it is well documented that fish stocks (especially reef fish stocks) are decreasing, it is not clear whether habitat degradation is the reason for this, or overfishing - which in turn contributes to habitat degradation. Probably it is a mixture of the two working together while also influencing each other, with such interactions very difficult to tease apart. Overfishing and/or habitat degradation can ultimately lead to collapsing fish stocks (pictured right).

Eutrophication – Tropical reefs are naturally low in nutrient levels (dissolved organic nitrogen & phosphorous for example). It is for this reason that over time corals have developed a symbiotic relationship with tiny dinoflagellate algae (zooxanthellae) that live within their tissue, an association that provides both partners with nutrients and other metabolites that facilitates their survival (page 39). Over recent years however nutrient levels have been increasing, a process known as eutrophication, which encourages the growth of algal species. Some of these species are planktonic and thus gradually change the crystal clear waters into a murkier green colour which blocks out light inhibiting photosynthesis in the coral's zooxanthellae. When this phytoplankton dies it adds to the water's sediment load which is again detrimental to coral as they easily become smothered. Furthermore these increased nutrients encourage large macroalgae to grow which can again smother corals, or new algal growths can outcompete new coral recruits. The effect of eutrophication can be accentuated by reducing the abundance of grazing creatures, which, in the case of herbivorous fish, can be caused through overfishing. Although increased nutrients can potentially increase the overall productivity of an area, providing food that may lead to an increased abundance of fish, a reef smothered by algae is an unhealthy ecological situation that will ultimately cause reef fish stock collapse through structural habitat degradation. Eutrophication is also a contributing factor in recent sargassum (page 16) beach inundations that threaten tourism and overall beach ecology. The source of nutrients may in part be from local sources (water outflows, leeching and/or runoff), but is in fact more likely a regional phenomenon with sources including (but not limited to) nutrient-rich river plumes from the Amazon and Orinoco, as a result of intensive farming and the continued deforestation of northern South America.

Dead Elkhorn Coral reef at Shoal Bay East with no fish present

Green water at Sile Bay signifies seasonal eutrophic conditions

Sargassum inundations are a sign of regional eutrophication

Algae covered dead Elkhorn Reef

Finger Coral being smothered by *Dictyota* sp. algae (page 17)

Bleached Low Relief Lettuce Coral (page 61)

Coastal development – If conducted responsibly then such activity can be carried out with little detriment to the marine (or terrestrial) environment. However, as is so often the case around the world this does not happen very often. For example by landscaping areas and/or removing beach flora, rainwater runoff into the sea is encouraged. This runoff is likely to have picked up fertilizers from the landscaped areas which will be carried unhindered to the sea. Increased sediment loads are also likely for similar reasons. Grey waters, or worse sewage, may also enter the ocean through careless developmental practices.

Disease – Over recent decades increased incidences of disease have been noticed among some marine creatures. Corals themselves have suffered a great deal of mortality though diseases such as white band disease and yellow blotch disease (page 69). The die off of extensive Elk Horn (*Acropora palmata*) and Stag Horn (*Acropora cervicornis*) 'forests', as pictured on page 220, combined with vast tracts of Orbicella Complex (*Orbicella* sp.) reef have been attributed to these two diseases, which are still present today. Another example of disease having profound effects on reef environments occurred during the early 1980s when Long-Spined Sea Urchins (*Diadema antillarum*) began to die off in Panama. The disease slowly spread and decimated populations throughout the entire Caribbean. These important grazers kept macroalgae in check, and today, although populations are eventually beginning to recover, it is feared that they are returning too late to be able to reverse the algal dominance that has taken hold since nutrient levels began increasing (pictured left top and middle).

Sea surface temperature rise – The much debated subject of climate change is one that is not necessary to get involved with here (for reading on the subject see page 228). What is known is that if sea temperatures increase much above 30°C and remain there for extended periods then corals bleach. Bleaching (page 70 and pictured left) is when a coral expels its zooxanthellae due to stress, and although it can recover, if it remains bleached for too long it will essentially die of starvation. Over recent decades the frequency of 'bleaching events' caused by periods of elevated sea surface temperature have increased. The last such event was in 2005, and reportedly caused widespread mortality to Caribbean reefs. If this frequency remains the same or increases then, coral bleaching looks set to be, along with eutrophication, one of the biggest threats facing coral reefs today.

Ocean acidification – Increased carbon dioxide in the atmosphere means more is absorbed by sea water, a chemical process that produces carboxylic acid which decreases the waters pH (acidification). Although the effect of this is only just beginning to be understood, and the severity currently minimal, it is widely agreed that this is a real threat to coral reefs as it will interfere with ocean chemistry and thus make it more problematic for coral species to lay down their calcium carbonate (alkaline) skeletons.

Physical damage – The sources of this problem are varied but all cause detriment to marine habitats that in turn contribute to habitat degradation. Boat groundings, anchor damage, standing or walking on the reef (amazingly this used to be considered an acceptable pastime for tourists), fish pot damage and/or net entanglement (both pictured right) all contribute to this physical disturbance. It only takes a slight touch to kill coral polyps or dislodge other creatures, and although such damage may seem insignificant it builds up over time, especially if other factors are inhibiting coral growth as discussed in some of the earlier points. This is especially poignant when considering hurricanes. This natural phenomenon has damaged reefs for millennia and yet they have survived. Today in Anguilla vast stretches that were shattered during the hurricanes of the 1980s and 1990s have not recovered. Hurricanes are thought to have once helped propagate reefs by spreading fragmented coral colonies around that re-attach and continue to grow once the storm has passed. It is thought that the stress of such damage, when combined with the added pressures corals face today, has inhibited such recovery. This is very bad news, especially when considering that hurricanes are predicted by some climate scientists to increase in the future due to climate change.

Invasive species – Because of the balance that exists in an ecosystem, a newly introduced species from another region can easily offset this. If an introduced species begins to flourish it can have a devastating effect on the native flora and fauna, and becomes known as an invasive species. Over recent years, the Indo-Pacific Lionfish (page 179) has begun to spread throughout the region. Being poisonous and a voracious predator of juvenile reef fish (thus potentially effecting future adult reef fish numbers), its arrival in 2010 was important news. The other invasive marine species in Anguilla at the current time are Orange Cup Coral (page 63), and a species of seagrass very similar in appearance to Midrib Seagrass (page 16).

Fishing gear can get tangled around coral heads

Fish traps can cause damage to live coral colonies

Garbage may be utilised by marine life such as this moray in a bottle

A plastic bag smothering a sea rod colony

Pollution – Aside from eutrophication mentioned earlier, and garbage (mentioned following), other human inputs into the marine ecosystem can produce environmental damage. Hydrocarbons that either leak from poorly maintained boats or are the result of accidental spills are a significant issue as not only can it harm sea birds and other creatures who live on the surface, but is can also interfere with the internal chemistry of creatures dwelling on the coast if washed up. Heavy metals, that may leech into the marine environment from terrestrial sources, are another cause for concern as levels can quickly become toxic, not only to marine life but to humans too. Also of recent interest are chemicals present in certain sunscreens that are reportedly toxic to corals. Some studies have shown that in areas were large numbers of tourists swim, corals can be negatively impacted.

Garbage – Leaving garbage lying around is an eyesore, but more importantly it can harm a great variety of marine life once it gets washed into the sea. Plastics are by far the biggest problem as relatively heavy inert materials such as glass and metal quickly sink and become encrusted by coralline algae or buried by sediment. Larger items can even offer fish and other creatures a place to live. Plastics however are a different story as the majority float and can be misleading for creatures such as turtles who mistake them for food and attempt to eat them. Consuming such items can cause their digestive systems to block and may ultimately lead to their death. This is especially the case for smaller plastic particles known as microplastics (see note on opposite page). Also, polythene can partially sink, but remain in the water column for extended periods and, aside from being mistaken as prey, it can become entangled around corals and smother them (left).

Unfortunately a great many of these threats have roots on a global scale and so very little can be done on a local level to make fundamental changes. To do so requires large-scale regional co-operation. That is not to say however that individual countries should sit back and wait for some kind of co-ordinated regional effort to be started before any local action is taken. On the contrary, it is very important that every nation, no matter how small, takes steps to protect and conserve not only the marine environment, but every environment that is being detrimentally effected by human activities. Often, many of these local environments can be effectively protected through local action, and because they are often the breadbasket of small island states, this protection is essential.

6.2 Marine Protection and Conservation Efforts in Anguilla

The marine environment is governed under a number of Acts in Anguilla including, but not limited to: The Fishery Protection Act; The Marine Parks Act; The Cruising Permits Act; and The Beach Protection Act. However, the legislative structure is continually evolving and so if of interest it is best to contact the Attorney Generals office to find out the most up-to-date situation. Aside from legislation and its enforcement, certain Government departments and other organisations play an important role in protecting the marine environment, as detailed below.

The Department of Fisheries and Marine Resources also heads a number of other initiatives whose long-term aim is the preservation of the marine environment. For example, the mooring buoy system, installed and maintained by the Department (page 8), is designed to deter boat users from anchoring in the Marine Parks. The Department also undertakes annual monitoring at fifteen coral reef and seagrass bed sites; monitors sand movement on many of Anguilla's beaches; conducts weekly in-water turtle surveys; and carries out regular research projects that seek to address various marine related issues. Such projects have included: the translocation of Long-Spined Sea Urchin populations from threatened habitats to 'safe houses' within Marine Parks; baseline study of sea water quality at eighty locations around the island; a population assessment of the invasive Lionfish; Spotted Spiny Lobster fishery assessment; Queen Conch fishery assessment; and fish diversity assessments on all of the wreck dive sites.

The Anguilla National Trust also conducts important work on Anguilla that contributes to the protection and conservation of marine ecosystems. The Trust has been involved with a number of important projects over recent years, including many conducted in partnership with the Department of Fisheries and Marine Resources that were focused on specific Marine Park areas. The Trust also helps to co-ordinate meetings of the Environmental Club, an educationally inspired organisation dedicated to instilling environmental awareness to young Anguillians. Usually in conjunction with local businesses, beach cleanups are organised from time-to-time that retrieve a shockingly large amount of debris. One such activity at Savannah Bay in 2008 collected over 100 black bin liners in a three and a half hour period. Other such efforts, for example on Blowing Point beach, collected somewhere between 30 and 50 bags.

The Department of Environment was established to help guide terrestrial based environmental issues, heading initiatives such as the Anguilla national ecosystem assessment. Over recent years however, they have also branched out to conduct project work in the marine environment, such as an underwater marine mapping project conducted in 2016/17. Moving on into the future, it will be necessary for all three agencies to work together in a coordinated fashion if the growing pressures on the waters surrounding Anguilla are to be successfully addressed and managed efficiently.

One of the most important conservation efforts to be achieved is on an individual scale. Unfortunately many people do not realise how the efforts of individuals can make a collective difference and think 'what is the point in me making an effort if nobody else does?'. Although the answer to this is often argued to be 'nothing', in actual fact many millions of people out there probably think the same thing, and if everybody decided to do what was best for the environment then a big difference would be achieved. Furthermore, by acting in a responsible way, one sets a good example that may inspire others to be more environmentally aware, especially the all-important younger generation.

What are microplastics? Although by definition microplastics are any plastic fragments less than 5 mm in diameter, and can be derived from larger plastic items that break up once discarded into nature, of particular concern are tiny manufactured particles that enter ecosystems directly. These include microfibres derived from clothing and microbeads (nurdles) from cosmetic products that wash down our drains. The presence of these sometimes microscopic plastic fragments in the marine environment has become of increasing interest over recent years. They can be ingested by small creatures (such as filter feeding bivalve species or grazing gastropods), and potentially work their way up the food chain. It is especially worrying that the toxic chemical compounds found within the plastics may leech out and over time slowly poison the consumer. Microplastics have been found even in 'pristine' environments such as the Pyrenees mountains, and concerns even exist in terms of airborne particulates.

6.3 Guidelines to Acting Responsibly

- **Snorkelling & diving** - Do not touch anything! "Leave only bubbles and take only memories" is a common slogan, and by doing so not only are you ensuring you cause no damage to the ocean realm, but you are also making sure you do not get spiked or stung by any of the creatures that live down there. Especially important is not to stand on rocks when snorkelling or kick them with your fins. Although they may look like rocks, in all likelihood they are living corals, sponges or other organisms that are highly susceptible to interference.

- **Fishing** – If partaking in fishing then it is conscientious behaviour to return juvenile fish back to the water. These smaller individuals have not yet had the chance to breed and so have not yet contributed to the next generation. Without their contribution it is more likely that fish stocks will diminish. Although these smaller fish are considered valuable bait to fishers, it is better to use the larger unwanted fish for this, or other bait sources i.e. cow-hide. Return to the water any individuals that appear to be breeding, and do not molest those individuals who look like they are exhibiting reproductive behaviour (see fish behaviour book in the reading list on page 228). It is also important not to throw fish traps directly onto the reef as they can cause considerable damage (as pictured on page 223). Doing so also increases the chance of their buoy lines becoming tangled around coral heads which not only damages them but also has the potential to sever the line and result in trap loss. Losing traps in such a way may contribute to 'Ghost Fishing', a process that refers to the ability of discarded or lost traps to continue fishing for a number of years after their loss unknown to their owners. To avoid this, easily degradable 'doors' can be inserted into the trap that rot after only a few weeks thus offering escape to otherwise doomed fish.

- **Boating** – If at all possible, it is best to avoid anchoring anywhere except on pure sand. Anchors cause great damage to seagrass beds and rocky areas that are covered with marine life, so if you are not sure of what is under the water then do not drop anchor. In a similar way it is also best not to navigate shallow areas unless you have knowledge of the underwater topography. Not only will a collision with the reef cause your boat substantial damage, but it will also damage the reef itself and the life that resides there. It is also important to use biodegradable cleaning products (detergents) while boating, and ensure all garbage is safely stowed away. This is because the majority of cleaning products, together with any loose garbage, will almost certainly end up in the ocean. The build up of detergents and other pollutants in the sea can adversely affect water quality, while garbage can sink and smother corals, or be mistaken for food by other creatures. Turtles, for example, have been documented with digestive systems blocked by plastic bags, cigarette butts, and other items. Finally, boat owners are reminded that it is an offense to discharge their waste water in any of the Marine Parks, and they are also asked to respectfully refrain from such activity anywhere else close to the coast or coral reef habitats.

- **Development** – Although most developers are large businesses, they are operated by individuals. If these individuals make responsible decisions then huge changes can be achieved. Sometimes these developers argue that making 'green' decisions can be costly and as such interfere with profits. This is often not the case, with some choices actually saving money. For example, to choose to turn your beach-front lights off at night, or at least dim them, not only saves money but also reduces energy consumption (pollution) and lighting disturbances to nesting turtles or their hatching babies. Another unfortunate activity that is related to irresponsible development is illegal sand mining. It is the practice of digging out sand from the beach or dune system to be sold and used in the construction industry. It can lead to changes in the erosional regimes around the island and increase runoff and sediment input into the ocean. Unfortunately it is a relatively common activity in Anguilla. Sile Bay, for example, used to have a beautiful beach and impressive dunes, although since their removal a few decades ago the area is still predominantly rocky. The same thing has been happening over recent years at Windward Point Bay (one of the eighteen protected beaches), with a similar end result. Removing sand from beaches is prohibited because it takes an unknown number of years (or decades) to return, and may never fully recover. Not only are the beaches important for Anguilla's economy, but they also protect the land behind, so their removal means the sea penetrates further inland causing damage to property. This is especially the case during storm surges, where the sea may breach a large distance inland if the beach and dunes have been removed. Any sand mining should be reported to the Royal Anguilla Police Force.

- **Living** – Even our day to day life away from the coast can have an impact on the sea. Some of the threats mentioned previously in this chapter are global in scale, and some, for example potential climate change, seem on far too much of a grand scale for any one person to do anything about. However, even simple acts such as turning off unwanted lights or switching the television off rather than leaving it on standby can make a huge difference. By saving energy less is needed to be produced and thus less pollution is released into the atmosphere. Less pollution in the atmosphere means a reduction in potential effects from climate change which may in turn reduce the frequency of bleaching events on coral reefs. In a similar manner sharing a car or reducing unnecessary journeys can have a collective effect if people make the right choices.

- **Beach going** – There are many ways in which the beach goer can act more responsibly, most of which have already been covered in this section or elsewhere in this book. As a final recap however it is a very good rule of thumb to leave no trace of your visit once you depart. This means clearing up any garbage that you may have left lying around and not removing any flora or fauna. Furthermore, if any garbage is present in your vicinity when you visit the beach, it is very diligent and beneficial behaviour to remove it when you depart. Because a great deal of material washes in from the ocean, it means that even if everyone takes all their waste away with them there will still be garbage on the beach.

6.4 Further Reading

The Reef Set (P. Humann & N. Deloach)

Reef Fish Behaviour (N. Deloach & P. Humann)

Guide to Marine Life: Caribbean – Bahamas – Florida (M. Snyderman & C. Wiseman)

Smithsonian Handbook of Whales Dolphins and Porpoises (M. Carwardine)

Sea Shells of the West Indies (M. Humfrey)

A Field Guide to Shells: Atlantic and Gulf Coasts and the West Indies (R. Tucker Abbott & P. A. Morris)

Marine Plants of the Caribbean (D. S. Littler, M. M. Littler, K. E. Bucher & J. N. Norris)

Global Marine Biological Diversity (E. A. Norse – Editor)

The Ecology of Reef Fishes (P. F. Sale)

Marine Ecology (R. S. K. Barnes & R. N. Hughs)

Life and Death of Coral Reefs (C. Birkeland – Editor)

Coral Disease Handbook, Guidelines for assessment, monitoring and management (L Raymundo, C. Couch & C.D. Harvell – Editors)

The Science and Politics of Global Climate Change (A. E. Dessler & E. A. Parson)

Responsible Tourism: Critical Issues for Conservation & Development (A. Spenceley)

6.5 Online Resources

Identification

Reef Guide – https://reefguide.org – A guide to the tropical fish of the world.

Coralpedia – https://coralpedia.bio.warwick.ac.uk – A guide to Caribbean corals and sponges.

Sponge Guide – http://www.spongeguide.org – A guide to Caribbean sponges.

Catalogues

Fish Base – https://www.fishbase.org – An online database of fish species of the world.

Algae Base – http://www.algaebase.org – An online database of algae species of the world.

Reef Base – http://www.reefbase.org – A global information system for coral reefs.

6.6 Glossary

Ahermatypic - Relating to a coral species that does not contribute to the overall build up of reef structure through deposition of calcium carbonate.

Barbel - A slender, whisker-like sensory organ near a fishes mouth. Usually on the under-side of a fishes head, and used to 'feels around' for food below the sand or sediment. Compare 'cirri'.

Benthic Zone - Relating to the bottom of a body of water.

Bioluminescence - The production and emission of light by a living organism as the result of a chemical reaction during which chemical energy is converted to light energy.

Caudal Peduncle - The narrow part of a fishes body where the tail attaches.

Chitin - A substance (long-chain polymer that is a derivative of glucose) which is the major constituent in the exoskeleton of arthropods.

Cirri - A slender hair-like structure similar to a tentacle, but generally lacking the tentacle's strength, flexibility and thickness. Usually on the upper-side of a fishes head, and used to sense changes in the water immediately above. Compare 'barbel'.

Cleaning Station - A location that is populated by cleaner fish and/or shrimps and visited by larger fish who allow themselves to be picked clean of ectoparasites.

Continental Shelf - The area of seabed around a land-mass where the sea is relatively shallow compared to the open ocean.

Corallite - The usually cup-shaped skeleton of an individual coral polyp, in which the polyp sits and into which it can retract.

Demersal - The part of the ocean close to the seabed. In terms of species, this refers to those that live and feed close or near to the bottom.

Dinoflagellate - Single-celled algae that live as phytoplankton but also form a symbiotic relationship with Cnidarians (most notably the corals) as Zooxanthellae.

Dorsal - The back or upper side of an organism.

Ectoparasites - Parasites that live on the external surfaces of other plants or animals.

Egg Trading - The ability of simultaneous hermaphrodites to swap sex rapidly and thus share the role of producing eggs which is an energetically expensive process. This then either reduces the energetic pressure placed on one of the mating pair, or allows double the amount of eggs to be produced.

Elasmobranch - Generic term for cartilagineous fish (sharks and rays)

Epifaunal - Those creatures that live on the outer surface of their environment, i.e. on the sea floor.

Epiphyte - A plant that grows on top of another plant or animal.

Eutrophic - Water that is rich in nutrients, or richer in nutrients than their natural state dictates.

Epifaunal - Refers to creatures that live on the outside of their environment, i.e. on or above the sea floor. Compare 'infaunal'.

Focus Biting - The habit of parrotfish to repeatedly bite particular areas of a coral head that leads to a concentration of damage to small isolated areas.

Gamete - An organisms reproductive cells.

Gorgonin - A complex protein that makes up the horny skeleton of the gorgonians (soft corals).

Hermatypic - Relating to a coral species that contributes to the overall build up of reef structure through deposition of calcium carbonate.

Infaunal - Refers to those creatures that live on the inside of their environment, i.e. under the sea floor. Compare 'epifaunal'.

Intertidal Zone - The area that is exposed to the air at low tide and submerged at high tide.

Keystone Species - A community member that plays a significant role in shaping an ecosystem.

Littoral Zone - Definitions differ, but in terms of usage in this book it refers to areas related to the shore. Thus it could extend from beyond any coastal sand dunes down into water of depths that are still effected by wave action etc.

Mantle - A highly significant part of molluscan anatomy that is usually responsible for shell formation. It has also been adapted for many other uses, for example, the siphon in many infaunal species; or the 'body' of a squid. In some gastropods the mantle extends out from the shell and surrounds it, acting as an effective camouflage.

Nematocyst - Specialised cells found in many marine invertebrates (most notably the Cnidarians), that contains a harpoon-like toxin-laden sting that is triggered by touch.

Ocellated - An eye-shaped spot, or spot of colour with differently shaded surrounding border.

Oligotrophic - Refers to water devoid of nutrients, or lacking the nutrients that their natural state dictates.

Operculum - The hard flap serving as a cover for either the gill slits of fishes, or the opening of the shell in certain gastropods when the body is retracted.

Oviparous - Refers to creatures that lay eggs.

Ovoviviparous - Refers to creatures who produce eggs that hatch inside the female. It differs from vivipary in that the young are nourished by a yolk sac rather than taking nourishment directly from the mother.

Pelagic - Deeper open-water, usually away from the continental shelf, rather than coastal or inland waters. In terms of species, this refers to those that live and feed in the open ocean without the need for a direct link to shallow areas.

Photosynthesis - The ability of plants (or other autotrophs) to generate carbohydrates and oxygen from carbon dioxide, water, and light energy using their chloroplasts.

Phytoplankton - Tiny, often single-celled plants and algae, that float freely in the water column. Zooplankton are their animal counterparts, many of which are larval stages that actively swim around and feed on phytoplankton before eventually settling and maturing on suitable habitat.

Polarised School - A group of fish all belonging to the same species that move in the same direction as each other simultaneously.

Protandrous Hermaphrodites - Fish or other creatures that change sex from male to female.

Protogynous Hermaphrodites - Fish or other creatures that change sex from female to male.

Recruiting/Recruitment - The process of acquiring new individuals (recruits) for a population through breeding.

Resident Spawning Aggregations - Refers to those spawning aggregations where partaking individuals live only a short distance away.

School - A group of fish, all often belonging to the same species, behaving in a uniform fashion.

Scute - A bony external plate or scale overlain with a horny substance.

Settling/settlement - Refers to the process when a free-swimming larva comes to rest on the sea floor or other structure, affixes itself, and takes up permanent residence.

Septa - A septum (plural septa) is one of a coral polyps radiating vertical plates lying within the corallite wall.

Sequential hermaphrodites - Fish or other creature that is born with the ability to change sex at some stage in its life history.

Shoal - A group of fish that do not all necessarily belong to the same species, and not all behaving in a uniform fashion.

Simultaneous hermaphrodites - A species with simultaneously functioning testes and ovaries.

Splash Zone - The part of the littoral zone that receives salt water via breaking waves.

Spawning Aggregation - A group of fish from the same species that come together at certain times of year to breed in unison.

Sublittoral Zone - Those areas beyond the littoral zone that are usually related to the continental shelf.

Swim Bladder - An internal gas-filled organ that helps many bony fish to control their buoyancy. This allows them to remain at a constant water depth without having to waste energy swimming.

Teleosts - Generic term for bony fish (jacks, mackerels, snappers, groupers etc).

Transient Spawning Aggregations – Refers to those spawning aggregations where partaking individuals live a considerable distance away, which thus involves the synchronisation of travel to the spawning site.

Ventral - The front or under-side of an organism.

Viviparous - Refers to creatures that give birth to live young.

Zooxanthellae - Single-celled dinoflagellate algae that live symbiotically in the tissue of many marine invertebrates, most notably the Cnidarians (corals etc). They provide nutrients to their host while in return consuming 'waste' metabolites.

Index of Common Names

F

G

H

Ribbon Worms 75
Rock Beauty 133
Rock Boring Urchin 100
Rock Hind 148
Roostertail Conch 85
Rorqual Whales 206
Rose Coral 60
Rose Lace Coral 41
Rosy Razorfish 166
Rough Box Crab 79
Rough Cactus Coral 62
Rough File clam 92
Roughlip Cardinalfish 171
Round Scad 113
Row Pore Rope Sponge 35
Ruby Brittle Star 97

S

Saddled Blenny 175
Sailfishes 108
Sailors Choice 128
Sand Diver 181
Sand Dollar 102
Sand Dollars 101
Sand Tilefish 184
Sargassum Triggerfish 190
Sargassum Weed 16
Saucereye Porgy 119
Saucer Leaf Alga 18
Saucer/Paddle Blade Alga 23
Saw Blade Alga 22
Sawcheek Cardinalfish 171
Scad 113
Scalloped Hammerhead 202
Scattered Pore Rope Sponge 35
Schoolmaster 132
Scorpionfishes 178
Scrawled Cowfish 187
Scrawled Filefish 191
Sculptured Slipper Lobster 78
Sea Anemones 45
Seabasses 144
Sea Biscuits 101
Sea Cucumbers 102
Sea Fans 68
Sea Frost 74
Sea Hares 90
Seahorses 182
Sea Pearl 23

Sea Plumes 67
Sea Rods 66
Sea Slugs 90
Sea Stars 95
Sea Turtles 206
Sea Urchins 98
Sea Walnut 71
Sea Whips 67
Segmented Worms 72
Sergeant Major 142
Sessile Barnacles 83
Sharksucker 197
Sharpnose Puffer 184
Sharptail Eel 199
Short-finned Pilot Whales 204
Shortfin Pipefish 182
Shrimps 75
Silversides 123
Sinuous Cactus Coral 62
Siphonophores 43
Six-Keyhole Sand Dollar 102
Skipjack Tuna 114
Slate-Pencil Urchin 101
Slender Filefish 192
Slimy Doris 91
Slippery Dick 164
Smallmouth Grunt 126
Smooth Flower Coral 63
Smooth Goose-Neck Barnacle 83
Smooth Star Coral 56
Smooth Trunkfish 188
Snake Eels 199
Snappers 129
Social Feather Duster 73
Soft Corals 65
Soldier Crabs 79
Solitary Disk Coral 62
Solitary Feather Dusters 73
Solitary Gorgonian Hydroid 43
Southern Sennet 117
Southern Stingray 202
Spadefishes 122
Spaghetti Worm 75
Spaghetti Worms 74
Spanish Grunt 128
Spanish Hogfish 161
Spanish Lobster 78
Spanish Mackerel 114
Speckled Cup Coral 64